# The Chemistry and Wonders of the Human Body

The Biochemic Statement of the Cause of Disease and the Physological and Chemical Operation of the Inorganic Salts of the Human Organism and Their Chemical Formulas; The Human Temple; The Chemical Bridge or Link Between Man and God.

By

DR. GEORGE W. CAREY

Author of Text Books on

The Chemistry of Life

Martino Publishing
Mansfield Centre, CT
2013

*Martino Publishing*
*P.O. Box 373,*
*Mansfield Centre, CT 06250 USA*

ISBN    978-1-61427-512-1

© *2013  Martino Publishing*

*All rights reserved. No new contribution to this publication may be reproduced, stored in a retrieval system, or transmitted, in any form or by any means, electronic, mechanical, photocopying, recording, or otherwise, without the prior permission of the Publisher.*

Cover design by T. Matarazzo

*Printed in the United States of America On 100% Acid-Free Paper*

# The Chemistry and Wonders of the Human Body

The Biochemic Statement of the Cause of Disease and the Physological and Chemical Operation of the Inorganic Salts of the Human Organism and Their Chemical Formulas; The Human Temple; The Chemical Bridge or Link Between Man and God.

By

### DR. GEORGE W. CAREY

Author of Text Books on
The Chemistry of Life

Published by
THE CHEMISTRY OF LIFE CO.
Los Angeles
1921

# TABLE OF CONTENTS

| | PAGE |
|---|---|
| The Chemistry of Life | 7 |
| The Chemistry of Wisdom | 9 |
| The Chemistry and Wonders of the Human Body | 11 |
| Disease, Nature's Efforts to Restore Equilibrium | 29 |
| Biochemistry | 31 |
| Fundamental Principles | 33 |
| Biochemistry and the Biochemic Pathology | 36 |
| The Chemistry of Blood and Tissue | 37 |
| Chemistry, Electricity, or Solar Energy | 40 |
| Cellular Pathology | 43 |
| The Views of Scientists | 45 |
| The Biochemic Pathology of Exudations, Swellings, etc. | 48 |
| Chemical Operation | 50 |
| Man's Divine Estate | 51 |
| Intuition, or the Infinite Vibration | 53 |
| The So-Called Elements | 55 |
| Esoteric Chemistry | 57 |
| The Fallacy of the Germ Theory of Disease | 60 |
| Diphtheria | 71 |
| Cholera | 73 |
| Pneumonia | 76 |
| Diabetes Mellitus | 78 |
| Syphilis | 81 |
| Gonorrhoea | 83 |
| Smallpox | 83 |
| Erysipelas | 85 |
| Intermittent Fever | 87 |
| Typhoid Fever | 90 |
| Yellow Fever | 92 |
| Lesson I | 94 |
| Lesson II | 97 |
| Lesson III | 99 |
| Lesson IV | 101 |

## Table of Contents

| | |
|---|---:|
| Lesson V | 104 |
| Lesson VI | 106 |
| Lesson VII | 109 |
| Lesson VIII | 111 |
| Lesson IX | 114 |
| Lesson X | 116 |
| Lesson XI | 118 |
| Lesson XII | 120 |
| Relation of the Mineral Salt of the Body to the Signs of the Zodiac | 121 |
| The Twelve Cell Salts of the Zodiac— | |
|     ARIES | 122 |
|     TAURUS | 124 |
|     GEMINI | 126 |
|     CANCER | 128 |
|     LEO | 130 |
|     VIRGO | 132 |
|     LIBRA | 134 |
|     SCORPIO | 136 |
|     SAGITTARIUS | 139 |
|     CAPRICORN | 141 |
|     AQUARIUS | 144 |
|     PISCES | 146 |
| Bioplasma, Life Substance | 148 |
| Diet | 149 |
| The Bridge of Life | 150 |

# THE CHEMISTRY OF LIFE

THE PERFECT HUMAN BODY IS THE RESULT OF A CERTAIN CHEMICAL FORMULA; DISTURB THE FORMULA AND SO-CALLED DISEASE RESULTS, FOR CHEMISTRY—GOD IN ACTION—IS NOT MOCKED.

When the arteries contain a sufficient quantity of the cell-salts, the aerial elements that form the organic portion of the blood are drawn into them by chemical affinity or magnetic attraction and precipitated or condensed to the consistency that forms the substance known as blood.

"The blood is the life."

The quality of blood depends entirely upon the chemical mineral base. If one or more of the inorganic salts are deficient in quantity, the blood will be deficient in vital or magnetic vibration, and cell and tissue building substance.

And so to supply the organism with the mineral principles that form the positive pole of blood is the natural, or physical, law of cure.

Lymph and the lymphatic system are parts of the complex, wonderful operation in the process of transmuting the etheric substance,—aerial elements,—into blood, flesh and bone.

The operation of wisdom has many names, but the chemical process is one.

The Chemistry of Life is God's creative compounds in action. It is the cosmic energy creating.

Life chemistry creates and changes its creations into multiple forms of infinite variety.

It is muck and lily, light and darkness, acid and alkali.

It is seen in the silver of the moon and the gold of the sun. It is heard in the cricket's plaintive song at eventide, and in the liquid melody of the mocking-bird that greets the morning light, and in the deep-toned thunder, when storm clouds meet. It is felt when its dynamic minerals

## The Chemistry and Wonders of the Human Body

chemicalize, and "Earth moves out of her place" and "Reels like a drunken man," and when the lava from some Vesuvius buries the "Cities of the plain."

It swings the Zodiac around the Seven Stars and hurls the comet on its way across the limitless reaches of star dust, leaving behind a billion miles of fire-embossed pathway.

This tremendous wizard is heard in the laugh of a child that strikes a finer chord of the human heart than tones from a harp strung with Apollo's beams when touched with the hands of an angel artist.

> Acid and Alkali acting,
>   Proceeding and acting again.
> Operating, transmuting, fomenting
>   In throes and spasms of pain—
> Uniting, reacting, creating,
>   Like souls "passing under the rod"—
> Some people call it Chemistry,
>   And others call it God.

# THE CHEMISTRY OF WISDOM
## A STATEMENT OF BEING

LIFE is omnipotent, omniscient and omnipresent. Being all, life must be wisdom. So, then, it follows, that all forms, appearances, and so-called matter must be life or wisdom in that form of expression.

"Ye are the salt of the earth, but if the salt have lost its savour, wherewith shall it be salted? It is thenceforth good for nothing but to be cast out and to be trodden under the feet of men."

"God made man from the dust (or mineral) of the earth."

"God made of *one* blood, all persons."

"The blood is the life."

"The life of the flesh is in the blood."

"Thou art Peter (Petra-stone or mineral), on thee will I build my church and the gates of hell shall not prevail against it."

The above words are simple statements of chemical and physiological facts.

The mineral salts of the human organism are intelligent entities, and work under divine guidance which man has designated as chemical affinity.

The molecules of iron, lime, magnesia, potash, etc., must have certain substances to work with, such as albumen, fibrin, oxygen, water, etc., or they *cannot* work, that is, "They have lost their savour," or substance.

The mineral salts are the base of the blood and good blood is the product of a proper balance of the dynamic molecules. Poor, or imperfect blood, is the product of a break in the molecular chain of salts. Having "lost the savour," some of the salts are good for nothing, etc., and hence the deficiency that causes so-called disease.

A thing is valuable only as it can be used. If the salt loses its savour, wherewith shall it be salted?

## The Chemistry and Wonders of the Human Body

The mineral salts may be present in the system, but we, through ignorance of nature's laws, may do that which will cause them to become of no value in the construction of the body. Not that the cell-salts themselves are changed in any way, but that the fluids of the body have become non-functional.

Science has proved that emotions of anger, fear, sorrow, etc., generate poisons in the human laboratory. Thought is the controller of the emotions. Therefore, our wrong thoughts, working through the emotions, have caused the fluids of the body to become poisoned or vitiated and thus thrown out of harmonious relation to the cell-salts, which cannot use non-functional oil, albumen, etc., in a manner to produce that harmonious condition called health.

This great chemical fact emphasizes the necessity of carefully guarding our thoughts, and makes plain the statement of Holy Writ: "As a man thinketh in his heart so is he."

Church, temple, house, etc., are derived from the Hebrew, Beth, and in the Scriptures these words are used to designate the human body. On petra, stone or mineral, the body is built.

Hell, hades, means stomach, and is derived from the fifteenth letter of the Hebrew alphabet, Samech.

Food is digested in the stomach and intestinal tract to furnish force, also to set free the mineral or cell-salts, and these remain *intact* and are *not in any manner changed;* therefore, "The gates of hell (grave or stomach) do not prevail against them."

The chemistry of the Old and New Testaments proves that the books were written by Masters.

The wisdom of the ages has produced none greater.

# THE CHEMISTRY AND WONDERS OF THE HUMAN BODY

> "For thou didst cover me in my mother's womb.
> I will give thanks unto thee; for I am fearfully and wonderfully made;
> Wonderful are thy works."
> —*139th Ps., 13 and 14th v.*

THE human race has been asleep, and has dreamed that property and money are the true wealth of a nation, sacrificing men, women and children to the chimerical idea that danced in visionary splendor through their brains. The result of this is to be seen in the uneasiness that prevails everywhere. But humanity is waking up, slowly but surely, and beginning to realize that it, itself, is the most precious thing on earth.

The old-established statement that the individuals that make up the race are imperfect is no more true than that a pile of lumber is imperfect, that is to be afterward re-formed, or built into a house. As it is the carpenter's business to take the lumber, which is perfect as material, and build the house, so it is the legitimate work of spiritual man to take the perfect material everywhere present and build, by the perfect law of chemistry and mathematics, the perfected, harmonious human being, and, with this material, employ the same law to build up society collectively.

It is a well-known physiological fact that the blood is the basic material of which the human body is continually built. As is the blood, so is the body; as is the body, so is the brain; as is the brain, so is the quality of thought. As a man is built, so thinks he.

According to the views of students of modern alchemy, the Bible—both the Old and the New Testaments—are symbolical writings, based primarily upon this very process of body building. The word alchemy really means fleshology. It is derived from chem, an ancient Egyptian word, meaning flesh. The word Egypt also means flesh, or anatomy.

## The Chemistry and Wonders of the Human Body

Alchemy, however, in its broader scope, means the science of solar rays. Gold may be traced to the sun's rays. The word gold means solar essence. The transmutation of gold does not mean the process of making gold, but does mean the process of changing gold, solar rays, into all manner of materialized forms, vegetable, mineral, etc. The ancient alchemist studied the process of Nature in her operations from the volatile to the fixed, the fluid to the solid, the essence to the substance, or the abstract to the concrete, all of which may be summed up in the changing of spirit into matter. In reality, the alchemist did not try to do anything. He simply tried to search out nature's processes in order that he might comprehend her marvelous operations.

To be sure, language was used that to us seems symbolical and often contradictory, but it was not so intended, nor true at all in reality. We speak in symbols. If a man is in delirium, caused by alcohol in his brain-cells, we say he has "snakes in his boots." Of course, no one supposes that the words are to be taken literally. Yet, if our civilization should be wiped out, and our literature translated after four or five thousand years, those who read our history might be puzzled to know what was meant by "snakes in his boots."

Again, it has been believed by most people that the words, "transmutation of base metals into gold," used by alchemists, referred to making gold. But a careful study of the Hebrew Cosmogony, and the Kabala, will reveal the fact that the alchemist always referred to solar rays when he used the word gold. By "base metals," they simply meant matter, or basic material. The dissolving or disintegration of matter, the combustion of wood or coal, seemed as wonderful to these philosophers as the growth of wood or the formation of coal or stone. So the transmutation of base metals into gold simply meant the process of changing the fixed into the volatile, or the dematerialization of matter, either by heat or chemical process.

It is believed by modern students of alchemy that the books of the Old and New Testament are a collection of alchemical and astrological writings, dealing entirely with the wonderful operation of aerial elements (spirit) in the human body, so fearfully and wonderfully made. The same

## The Chemistry of Wisdom

authority is given for the statement, "Know ye not that your bodies are the temple of the living God" and "Come unto Me all ye that labor and are heavy laden and I will give you rest." According to the method of reading the numerical value of letters by the Kabala, M and E figure B, when united. Our B is from the Hebrew Beth, meaning a house or temple—the temple of the spiritual ego—the body. Thus by coming into the realization that the body is really the Father's House, temple of God, the spirit secures peace and contentment or rest.

The human body is composed of perfect principles, gases, minerals, molecules, or atoms; but these builders of flesh and bone are not always properly adjusted. The planks or bricks used in building houses may be endlessly diversified in arrangement, and yet be perfect material.

Solomon's temple is an allegory of man's temple—the human organism. This house is built (always being built) "without the sound of saw or hammer."

The real Ego manifests in a house, beth, church, or temple—i. e., Soul-of-Man's Temple, for the Ego or I AM.

The solar (soular) plexus is the great central sun or dynamo on which the Subconscious Mind (another name for God) operates and causes the concept of individual consciousness. Specifically stated thus:

1. The upper brain (cerebrum); "The Most High" or Universal Father, which furnishes substance for all functions that constitute the body.

2. The Spiritual Ego ("I AM") resident in the cerebellum.

3. The Son of God, the redeeming seed or Jesus, born monthly in the solar plexus.

4. Soul, the fluids of the body.

5. Flesh, bone, etc., the fluids materialized. (In a broader sense body also is termed soul, "Every soul perished." It is not thinkable that every Spirit, or Ego or "I AM" was drowned.

No wonder that the seers and alchemists of old declared that "Your bodies are the temple of the living God" and, "The kingdom of Heaven is within you." But man, blinded by selfishness, searches here and there, scours the heavens

## The Chemistry and Wonders of the Human Body

with his telescope, digs deep into earth, and dives into ocean's depths, in a vain search for the elixir of life that may be found between the soles of his feet and the crown of his head. Really our human body is a miracle of mechanism. No work of man can compare with it in accuracy of its process and the simplicity of its laws.

At maturity, the human skeleton contains about 165 bones, so delicately and perfectly adjusted that science has despaired of ever imitating it. The muscles are about 500 in number; length of alimentary canal, 32 feet; amount of blood in average adult, 30 pounds, or one-fifth the weight of the body; the heart is six inches in length and four inches in diameter, and beats seventy times per minute, 4200 times per hour, 100,800 per day, 36,720,000 per year. At each beat, two and one-half ounces of blood are thrown out of it, 175 ounces per minute, 656 pounds per hour, or about eight tons per day.

All the blood in the body passes through the heart every three minutes; and during seventy years it lifts 270,000,000 tons of blood.

The lungs contain about one gallon of air at their usual degree of inflation. We breathe, on an average, 1200 breaths per hour; inhale 600 gallons of air, or 24,000 gallons daily.

The aggregate surface of the air-cells of the lungs exceeds 20,000 square inches, an area nearly equal to that of a room twelve feet square. The average weight of the brain of an adult is three pounds, eight ounces; the average female brain, two pounds, four ounces. The convolutions of a woman's brain cells and tissues are finer and more delicate in fibre and mechanism, which evidently accounts for the intuition of women. It would appear that the difference in the convolutions and fineness of tissue in brain matter is responsible for the degrees of consciousness called reason and intuition.

The nerves are all connected with the brain directly, or by the spinal marrow, but nerves receive their sustenance from the blood, and their motive power from the solar plexus dynamo. The nerves, together with the branches and minute ramifications, probably exceed ten millions in number, forming a bodyguard outnumbering the mightiest army ever marshalled.

The skin is composed of three layers, and varies from one-eighth to one-quarter of an inch in thickness. The average area of skin is estimated to be about 2000 square inches. The atmospheric pressure, being fourteen pounds to the square inch, a person of medium size is subject to a pressure of 28,000 pounds. Each square inch of skin contains 3500 sweat tubes, or perspiratory pores (each of which may be likened to a little drain tile) one-fourth of an inch in length, making an aggregate length of the entire surface of the body 201,166 feet, or a tile for draining the body nearly forty miles in length.

Our body takes in an average of five and a half pounds of food and drink each day, which amounts to one ton of solid and liquid nuorishment annually, so that in seventy years a man eats and drinks 1000 times his own weight.

There is not known in all the realms of architecture or mechanics one little device which is not found in the human organism. The pulley, the lever, the inclined plane, the hinge, the "universal joint," tubes and trap-doors; the scissors, grind-stone, whip, arch, girders, filters, valves, bellows, pump, camera, and Aeolian harp; and irrigation plant, telegraph and telephone systems—all these and a hundred other devices which man thinks he has invented, but which have only been telegraphed to the brain from the Solar Plexus (cosmic centre) and crudely copied or manifested on the objective canvas.

No arch ever made by man is as perfect as the arch formed by the upper ends of the two legs and the pelvis to support the weight of the trunk. No palace or cathedral ever built has been provided with such a perfect system of arches and girders.

No waterway on earth is so complete, so commodious, or so populous as that wonderful river of life, the "Stream of Blood." The violin, the trumpet, the harp, the grand organ, and all the other musical instruments, are mere counterfeits of the human voice.

Man has tried in vain to duplicate the hinges of the knee,

elbow, fingers and toes, although they are a part of his own body.

Another marvel of the human body is the self-regulation process by which nature keeps the temperature in health at 98 degrees. Whether in India, with the temperature at 130 degrees, or in the arctic regions, where the records show 120 degrees below the freezing point, the temperature of the body remains the same, practically steady at 98 degrees, despite the extreme to which it is subjected.

It was said that "All roads lead to Rome." Modern science has discovered that all roads of real knowledge lead to the human body. The human body is an epitome of the universe; and when man turns within the mighty searchings of reason and investigation that he has so long used without—the "New Heaven and Earth" will appear.

While it is true that flesh is made by a precipitation of blood, it is not true that blood is made from food. The inorganic or cell-salts contained in food are set free by the process of combustion or digestion, and carried into the circulation through the delicate absorbent tubes of the mucous membrane of stomach and intestines. Air, or Spirit, breathed into the lungs, enters the arteries (air carriers) and chemically unites with the mineral base, and by a wonderful transformation creates flesh, bone, hair, nails, and all the fluids of the body.

On the rock (Peter or Petra, meaning stone) of the mineral salts is the human structure built, and the grave, stomach, or hell shall not prevail against it. The minerals in the body do not disintegrate or rot in the grave.

The fats, albumen, fibrine, etc., that compose the organic part of food, are burned up in the process of digestion and transposed into energy or force to run the human battery. Blood is made from air; thus all nations that dwell on earth are of one blood, for all breathe one air. The best food is the food that burns up the quickest and easiest; that is, with the least friction in the human furnace.

The sexual functions of man and woman, the holy operation of creative energy manifested in male and female; the formation of life germs in ovum and sex fluids; the Divine

## The Chemistry of Wisdom

Procedure of the "word made flesh" and the mysteries of conception and birth are the despair of science.

"Know ye not that your bodies are the temple of the living God?" for, "God breathed into man the breath of life."

In the words of Epictetus, "Unhappy man, thou bearest a god with thee, and knoweth it not."

Walt Whitman sings:

"I loaf and invite my soul; I lean and loaf at my ease, observing a spear of summer grass. Clean and sweet is my soul, and clean and sweet is all that is not my soul."

"Welcome every organ and attribute of me, and of any man hearty and clean, not an inch, not a particle of an inch, is vile, and none shall be less familiar than the rest."

"Divine am I, inside and out, and I make holy whatever I touch or am touched from."

"I say no man has ever yet been half devout enough; none has ever yet adored or worshipped half enough; none has begun to think how divine he himself is, and how certain the future is."

The vagus nerve, so named because of its wandering (vagrant) branches, is the greatest marvel of the human organism. Grief depresses the circulation through the vagus, a condition of malnutrition follows, and tuberculosis, often of the hasty type, follows.

The roots of the vagus nerve are in the medulla oblongata, at the base of the small brain or cerebellum, and explains why death follows the severing of the medulla. It controls the heart action, and if a drug such as aconite be administered, even in small doses, its effect upon this nerve is shown in slowing the action of the heart and decreasing the blood pressure. In larger doses it paralyzes the ends of the vagus in the heart, so that the pulse becomes suddenly very rapid and at the same time irregular. Branches of the vagus nerve reach the heart, lungs, stomach, liver and kidneys.

Worry brings on kidney disease, but it is the vagus nerve, and especially that branch running to the kidneys which undue excitement or worry, or strain, brings about the paralysis of the kidneys in the performance of their functions.

## The Chemistry and Wonders of the Human Body

When we say that a man's heart sinks within him from fear or apprehension, it is shown by the effect of this nerve upon the heart action. If his heart beats high with hopes, or he sighs for relief, it is the vagus nerve that has conducted the mental state to the heart and accelerated its action or caused that spasmodic action of the lungs which we call a sigh.

The nerves of the human body constitute the "Tree of Life," with its leaves of healing. The flowing waters of the "Rivers of Life" are the veins and arteries through which sweep the red, magnetic currents of Love—of Spirit made visible.

Behold the divine telegraph system, the million nerve wires running through the wondrous temple, the temple not made with hands, the temple made "without sound of saw or hammer." View the Central Sun of the human system—the Solar Plexus—vibrating life abundantly.

Around this dynamo of God, you may see the Beasts that worship before the Throne day and night saying, "Holy holy, art Thou, Lord God Almighty." The Beasts are the twelve plexuses of nerve centers, telegraph stations, like unto the twelve zodiacal signs that join hands in a fraternal circle across the gulf of space.

Aviation, liquified air, deep breathing for physical development and the healing of divers diseases rule the day. In every brain there are dormant cells, waiting for the "coming" of the bridegroom, the vibration of the air age (the Christ) that will resurrect them.

Everywhere we have evidence of the awakening of dormant brain cells. Much, if not all, of spiritual phenomena, multiple personality, mental telepathy, and kindred manifestations are explainable upon the hypothesis of the possibility of awakening and bringing into use of dormant brain cells.

Scientists have discovered that there are dormant, or undeveloped brain cells in countless number, especially in the cerebrum, or upper brain, the seat of the moral faculties; or, more definitely speaking, the key, which, when touched with the vital force set free through the process of physical regeneration by saving the seed and by the baptism in ointment (Christ) in the spinal cord, lifts the crucified substance

## The Chemistry of Wisdom

wasted by the prodigal son in "riotous living" up to the "most high" brain.

This procedure causes the dormant cells, little buds, to *actually bloom*. The simile is perfect. The cells, while dormant, are like a flower yet in the bud. When the Substance that is needed for their development reaches them, through physical regeneration as outlined in the "Plan of Salvation," fully explained in God-man, the cells *bloom* and then vibrate at a rate that causes the Consciousness of the "New Birth."

"He that is born of God doth not sin, *for his seed remaineth in him.*" John.

The eye is hardly less wonderful, being a perfect photographer's camera. The retina is the dry plate on which are focused all objects by means of the crystalline lens. The cavity behind this lens is the shutter. The eyelid is the drop shuttle. The draping of the optical dark room is the only black membrane in the entire body. The miniature camera is self-focusing, self-loading and self-developing, and takes millions of pictures every day in colors and enlarged to life size.

Charts have been prepared which show that the eye has 729 distinct expressions conveying as many distinct shades of meaning.

The power of color perception is overwhelming. To perceive red the retina of the eye must receive 395,000 vibrations in a second; for violet it must respond to 790,-000,000. In our waking moments our eyes are bombarded every minute by at least 600,000,000 vibrations.

The ear contains a perfect miniature piano of about 3000 double fibers or strings stretched or relaxed in unison with exterior sounds. The longest cord of this marvelous instrument is one-fifteenth of an inch, while the shortest is about one-five-hundredth of an inch. The 3000 strings are distributed through a register of seven octaves, each octave corresponding to about 500 fibers and every half tone subdivided again into 320 others. The deepest tone we can hear has thirty-two vibrations a second, the highest has 70,000.

The ear is a colossal mystery, and the phenomenon of

sound is a secret only recorded in the Holy of Holies of the Infinite Mind. And what is mind? We know absolutely nothing about it. Some believe that mind is the product of the chemical operation of matter, viz.: the atoms or materials that compose the human body. These persons contend that all electrons are particles of pure Intelligence and KNOW what to do. Others hold to the theory that universal Mind (whatever it may be) forms a body from some material, they know not what, and then plays upon it or operates through it.

    Visions of beauty and splendor,
      Forms of a long-lost race,
    Sounds of voices and faces
      From the fourth dimensions of space;
    And on through the universe boundless,
      Our thoughts go, lightning-shod;
    Some call it Imagination,
      And others call it God!

Now comes speech, the Word that was in the beginning. God certainly bankrupted His infinite series of miracles when He gave the power of speech to man.

We wonder and adore in the presence of that pulsing orb, the heart. Tons of the water of life made red by the Chemistry of Love sweep through this central throne every day, and flow on to enrich the Edenic Garden until its waste places shall bloom and blossom as the rose.

Take my hand and go with me to the home of the Spiritual Ego—the wondrous brain. Can you count the whirling, electric, vibrating cells? No, not until you can count the sand grains on the ocean's shore. These rainbow-hued cells are the keys that the fingers of the soul strike to play its part in the Symphony of the Spheres.

At last we have seen the "Travail of the Soul and are satisfied." No more temples of the Magi now, but instead the glorious human Beth. At last we have found the true church of God, the human body. In this body, or church, spirit operates like some wizard chemist or electrician. No more searching through India's jungles or scaling the Himalayan heights in search of a master — a mahatma — or ancient priest, dwelling in some mysterious cave where occult rites and ceremonies are supposed to reveal the wisdom of the past. But, instead, you have found the Kingdom of the Real within the Temple that needs no outer Sun by day nor Moon or Stars by night to lighten it. And then the enraptured Soul becomes conscious that the stone has been rolled away from the door of material concept where it has slept, and it now hears the voice of the Father within saying, "Let there be light!" and feels the freedom that comes with knowing that Being is one.

And now man also realizes the meaning of the "Day of Judgment." He realizes that Judgment means understanding, hence the ability to judge. He then judges correctly, for he sees the Wisdom of Infinite Life in all men, in all things, all events, and all environments. Thus does the new birth take place, and the Kingdom of Harmony reigns now.

Man must realize, however, that he is the creator or builder of his own body, and that he is responsible for every moment of its building, and every hour of its care. He alone can select and put together the materials provided by the universe for its construction. Man has been able to scale the heavens, to measure the distance between the stars and planetary bodies, and to analyze the component parts of suns and worlds, yet he cannot eat without making himself ill; he can fore-tell eclipses and tides for years in advance, but cannot look far enough ahead in his own affairs to say when he may be brought down with la grippe, or to calculate accurately the end of any bodily ill that may afflict him. When he finds out what he really is, and how much he has always had to do in the making of himself what he is, he will be ready to grasp some idea of the wonderful possibilities of every human body, and will know how completely and entirely is every man his own savior. Just so long as he denies his own powers, and looks outside of himself for salvation from present or future ills, he is indeed a lost creature. If the race is to be redeemed, it must come as the result of thought followed by action. If the race is to think differently than at present, it must have new bodies with new brains.

Modern physiologists know that our bodies are completely made over every year, by the throwing off of worn-out cells and the formation of new ones, that is going on every minute. Nature will take care of the making-over process, but we are responsible for the plan of reconstruction. Man must learn to run the machinery of his body with the same mathematical accuracy as he now displays in control of an engine or automobile, before he can lay claim to his divine heritage and proclaim himself master of his own.

The law of life is not a separate agent working independently of mankind and separate from individual life. Man himself is a phase of the great law in operation. When he once fully awakens to the universal co-operation of the attributes and thoughts through which the great dynamo operates or proceeds, he—an Ego—one of the expressions of Infinity, will be enabled to free himself from the seeming environments of matter, and thus realizing his power, will assert his dominion over all he has been an agent in creating. And he has indeed assisted in creating—manifesting—all that is. Being a thought, an outbreathing of universal spirit, he is co-eternal with it.

In material concept, we do not begin to realize the extent of our wisdom. When we awaken to Ego, or spirit, consciousness—knowledge that we are Egos that have bodies or temples, and not bodies that have Egos—we see the object or reason of all symbols or manifestation, and begin to realize our own power over all created things.

And in this Aquarian age, great changes in nature's laws will be speedily brought to pass, and great changes in the affairs of humanity will result. The laws of vibration will be mastered, and through their operation material manifestations will be shaped and moulded to man's will. It is only a matter of time when all the necessities of life will be produced directly from the elements of the air.

It is well known by chemists that all manner of fruits, grains and vegetables are produced directly from the elements in the air, and not from the soil. The earth, of course, serves as a negative pole and furnishes the mineral salts of lime, magnesium, iron, potassium, sodium and silica, which act as carriers of water, oil, sugar, albumen, etc., and are formed by a precipitation or condensation of principles in the air, and not from the soil. This is a fact abundantly proved.

## The Chemistry of Wisdom

M. Berthelot, a scientist of France, Tesla, the Austrian wizard, and our own Edison have long held that food can be produced artificially by a synthetic process from its elements. Professor W. O. Atwater, Ph. D., of Wesleyan University, shows, in Farmer's Bulletin No. 23, how all vegetable growth, or substance, is 98 per cent condensed aerial elements and 2 per cent inorganic salts. He says: "Ash: The mineral matter, or ash, which is left behind when animal or vegetable matter is burned, consists of a variety of chemical compounds commonly called salts, and includes phosphates, sulphates and chlorides of the metals, calcium, magnesium, potassium and sodium. Calcium phosphate, or phosphate of lime, is the chief mineral constituent of bone. Common salt is chloride of sodium."

Again: "The potential energy of the food is transformed in the body into heat and mechanical power. Thus, in that respect, the steam engine and the body are alike."

"The starch of bread, potatoes, and sugar are burned in the body to yield heat and power."

"The fats of meat and butter serve the same purpose, only they are a more concentrated *fuel* than the carbohydrates."

Of course, Professor Atwater is here writing on the nutritive value of foods, and not to prove that *they are not transmuted into blood;* but, nevertheless, he does prove that very fact.

Some six or seven extracts, as well as coloring material are now being manufactured in this manner. Madder is made almost exclusively by this process now.

M. Berthelot, at one time the French minister of foreign affairs, possesses fame apart altogether from his political efforts. In his special domain of chemical knowledge he ranks among the first of his contemporaries. Chemical synthesis—the science of artificially putting organized bodies together—may be said to owe its existence to him. The practical results expected to flow from his experiments and discoveries are enormous. Thus, sugar has recently been made in the laboratory from glycerine, which Professor Berthelot first made direct from synthetic alcohol. Commerce has now taken up the question; and an invention has

## The Chemistry and Wonders of the Human Body

recently been patented by which sugar is to be made upon a commercial scale, from two gases, at something like 1 cent per pound. M. Berthelot declares he has not the slightest doubt that sugar will eventually be manufactured on a large scale synthetically, and that the culture of sugar cane and beet root will be abandoned, because they have ceased to pay.

The chemical advantages promised by M. Berthelot to future generations are marvelous. He cites the case of alzarin, a compound whose synthetic manufacture by chemists has destroyed a great agricultural industry. It is the essential commercial principle of the madder root, which was once used in dyeing, wherever any dyeing was going on. The chemists have now succeeded in making pure indigo direct from its elements, and it will soon be a commercial product. Then the indigo field, like the madder fields, will be abandoned, industrial laboratories having usurped their place.

But these scientific wonders do not stop here. Tea and coffee are to be made artificially; not only this, but there is substantial promise that such teas and such coffees as the world has never seen will be the outcome. Theobromine, the essential principle of cocoa, has been produced in the laboratory, thus synthetic chemistry is getting ready to furnish the three great non-alcoholic beverages now in general use.

Biochemists long ago advanced the theory that animal tissue is formed from the air inhaled, and not from food. The food, of course, serves its purpose; it acts as the negative pole, as does the earth to plant and vegetable life, and also furnishes the inorganic salts, the workers that carry on the chemistry of life, setting free magnetism, heat and electric forces by disintegration and fermentation of the organic portions of the food.

But air, in passing through the various avenues and complex structure of the human organism, changes, condenses, solidifies, until it is finally deposited as flesh and bone. From this established scientific truth, it appears that, by constructing a set of tubes, pumps, etc., resembling the circulatory system, as well as the lung cells of the human mechanism, which is a chemical laboratory, where the chemistry of spirit is ever at work, changing the one essence of spirit substance to blood, flesh and bone, air may be changed into an albuminous pabulum, which may be again changed into the special kind of food required by adding the proper flavor, which may also be produced direct from the air.

There does not seem to be any reason why this substance, the basis of all food or vegetable growth, cannot, by drying and proper process, be made into material for clothing. Wool, cotton, flax, silks, etc., are all produced from the universal elements through the slow, laborious and costly process of animal or vegetable growth. Why not produce them directly?

The St. Louis Post-Dispatch of August 31, 1902, contained an article under headlines as follows: "How Food May Be Made From the Air We Breathe," "The Latest Marvel of Science," "Atmospheric Food Company Incorporated at Niagara Falls."

"The interesting article in the Sunday Post-Dispatch describing how modern invention will enable us to draw new food supplies from the air itself showed how an up-to-date newspaper keeps its readers informed on the latest scientific discoveries.

"Readers of the article know that science can now produce nitric acid by burning the air by means of powerful electric arcs. The acid is made cheaper than it has heretofore been obtained by chemical action. And this cheap production means cheap nitrates. Cheap nitrates means more bushels of wheat per acre, and an increase in all kinds of crops. Peru and Chile went to war for the possession of beds of nitrates. Now we can obtain these fertilizers anywhere, by setting an electric current to work. The possibilities are stupendous.

"This gas, nitrogen, which makes four-fifths of the atmosphere, was a puzzle to early science. It seemed dead and useless. It was looked upon merely as a filler, to prevent the more active oxygen from burning the earth up. It appears that it may yet become an exhaustless common bank, from which humanity may draw wealth for all time. The story of its exploitation is one of the fairy tales of science."

Those who believe in a time of peace on earth, a millennial reign, certainly do not think that our present mode of producing food will continue during that age. Slaughter of animals, and fruit, grain or vegetable raising, leave small time for men and women to enjoy a condition foretold by all the seers and prophets. But, under the new way of producing food and clothing, the millennium is possible.

And thus will the problem of subsistence be solved. No more monopoly of nature's bounties. An exchange of service will be the coin of the world instead of certain metals difficult to obtain.

A realization of this vision, or theory, that will for awhile be called visionary by most people, will mean Eden restored. Many people have wondered why, during the last few years, fruit pests have multiplied so alarmingly, and why cows are almost universally diseased and so much attention given to meat, milk, and butter products by Boards of Health, etc. There is surely a reason for all this. The One Life, Supreme Intelligence, or Divine Wisdom, that holds the worlds of space in their appointed orbits, surely knows all about the affairs of earth. When a new dispensation is about to be ushered in, old things begin to pass away.

All labor of preparing food and clothing, as now carried on, will cease, and the people, in governmental or collective capacity, will manufacture and distribute all manner of food and clothing free. Machinery for the production of everything necessary for man's material wants will be simple and easily manipulated. One-twentieth of the able-bodied population, working one or two hours a day, and shifting every week, or day, for that matter, with others, will produce an abundant supply. Neither droughts nor floods nor winter's snow can affect the supply. It can be made in Klondyke or the Tropics. Garments may be worn for a few days and then burned, and laundry work cease. Cooking will be reduced to a minimum, as the food will only need flavoring. No preparing vegetables, fruits, or cracking nuts; no making butter, or preserving meats. Men will not have to devote their lives to the endless grind of food

## The Chemistry of Wisdom

production, nor woman to cooking, dish-washing, sewing, and laundry-work. Garments of beautiful design and finest texture will be made by machines invented for the purpose, ready for wear.

A dream, you say? I cannot admit that, in the face of the indisputable evidence it has already been able to produce, but what, if it were, at present, but the dream it may appear to be to the one who hears of its methods of operation for the first time, it is certain of future fulfillment. Do dreams ever come true? Yea, verily! All concrete facts are materialized dreams.

An Egyptian King dreamed, and the Pyramids of Cheops mass and miracle his vision. The Pyramids are encyclopedic of physical science and astral lore. The science of numbers, weights, measures, geometry, astronomy, astrology, and all the deeper mysteries of the human body and soul are symboled in these incomparable monuments.

A dream of an ancient alchemist solidified in stone, and the awful sphinx sat down in Egypt's sand to gaze into eternity.

Columbus dreamed, and a white-sailed ship turned its prow west and west. On uncharted seas, with an eternity of water ahead, he remembered his dream, and answered "Sail on!" to the discouraged mate, until he landed on the unknown shores of a most wonderful new world.

Michael Angelo dreamed a thousand dreams and sleeping marble awoke and smiled. Hudson and Fulton dreamed, and steamboats "run over and under the seas."

The Pilgrim Fathers dreamed and America, the "marvel of nations," banners the skies with the stars and stripes. Marcus Whitman and Lewis and Clarke dreamed long and hard and the bones of oxen and men and women and babies made a bridge over the desert sands and the mountain gorges to the shores of the Sundown Sea, and now the Pullman cars come safely over. Morse and Marconi and Edison dreamed strange wild dreams and concentrated intelligence springs from carbon-crucible and says to earth's boundaries, "Lo, here am I."

## The Chemistry and Wonders of the Human Body

Vibration of etheric substance
   Causing light through regions of space,
A girdle of something enfolding
   And binding together the race—
And words without wires transmitted,
   Aerial-winged, spirit-sandalled and shod:
Some call it electricity,
   And others call it God!

A mechanic dreamed, and sprang upon his automobile, and drove it till the axles blazed and the spaces shriveled behind him. Men of high strung airy brains dreamed wondrous dreams, and now the eagle's highway and the open road of men lie parallel.

A musician dreamed a sweet, harmonious dream, and forth from a throat of brass directed by a million tiny fingers of steel, came the entrancing notes that have run riot through the singer's brain.

So let us dream on, men and women, of the day of rest that is already dawning in the heavens. No wonder that Paul said, "Now, brethren, are we the Sons of God, but it doth not yet appear what we shall be,"—as such. The morning light of that glad day now purples the mountains of faith and hope with its rays of glory.

And when Man is once fully alive to his own heritage, realizing the wonders and possibilities of his own body, and the power of his spirit to control it, and to provide for its needs, he will assert the divine right within him to be a spirit in command of its own temple, and the environment of that temple, and will rejoice in the revealed truth of his own divinity that alone can make him free.

# DISEASE NATURE'S EFFORTS TO RESTORE EQUILIBRIUM

DISEASE is an alarm signal, a friend who calls to inform us of danger. Disease is an *effort to prevent death*.

Therefore, pain and so-called disease is more than a warning; it is an effort that opposes death. The symptoms that indicate disease are calls, or dispatches, asking for the material with which the repair of bodily tissue may be made. Pains or discomforts of various functions or structure of the body are *words* asking for the constituent parts of blood, nerve fluids, tissue, bone, etc.

If acids cause pain, the pain is a call for a sufficient amount of alkaloid salts to counteract an acid effect and change fluids to a bland and natural state.

Healthy synovial fluid (fluids of the joints) is neither acid nor alkali, but yet contains both in combination. Should the alkaline salts become deficient in amount, for any cause, the acid at once becomes a disturbing element and hurts the nerves that pervade the membranes of the periosteum (bone covering) of the internal structure of the knee, elbow or other joints of the human anatomy. This pain, or word, can not be considered bad or malignant in any sense.

All phenomena appears as a result of Divine, beneficent law, hence disease so-called is the result of the orderly procedure of that law. In all ages men and women have been sick more or less. In all ages these have been storms, cataclysms, earthquakes and extremes of heat and cold; no one questions the wisdom that causes, guides and directs these events, then why should we question the wisdom of disease? Disease is one phase of the transmutation of matter in the procedure of regeneration.

All methods of healing are phases of the transmutation process.

## The Chemistry and Wonders of the Human Body

When man reaches the plane of understanding (Alchemical Knowledge) he will consciously co-operate with the *Divine Urge by supplying his* dynamic laboratory with the mineral base of the blood (the *Philosopher's Stone*), and thus make blood the "Elixir of Life" as it is destined to be.

Consider Biochemistry, thou invalid, study her truths, practice her precepts, and you will obtain wisdom, and realize you are a worker in the plan of regeneration.

# BIOCHEMISTRY
## "THE STONE THE BUILDERS REJECTED"

"I know perfectly well my own egotism;
I know my omnivorous lines,
And will not write any less,
And would fetch you, whoever you are, flush with myself."
—*Walt Whitman.*

THE constituent parts of man's body are perfect principles, but the principles are not always perfectly adjusted.

The planks, bricks, or stones with which a building is to be erected are composed of perfect principles, namely, oxygen, hydrogen, carbon, lime, iron, silica, potassium, magnesia, etc. These principles or elements are eternally perfect *per se*, but may be endlessly diversified in combination.

The stone which the builders rejected is symbolized by the stone which the builders of the pyramid of Cheops failed to place in position on the *top corner*—the pyramid being five-cornered, one corner pointing upward, and representing the sense of seeing—so the builders of the science of medicine have failed to place the mineral basis of blood—the inorganic salts—in their place in the human structure or fleshly pyramid.

When these mineral (stone) principles, or elements, are perfectly placed in the chemical formulae which composes the blood, the animal functions proceed in harmonious operation. When for any reason these cell-salts, stones, are deficient or negative or dormant or get misplaced, i. e., out of combination, the stone which must become the head of the corner has been rejected by the chemistry of life builders.

The human body, or pyramid, is a storage battery, and must be supplied constantly with the proper elements—chemicals—to set up motion or vibration at a rate that will produce what we please to call a live body. A failure to

## The Chemistry and Wonders of the Human Body

keep the storage battery supplied with the chemical base of blood causes a disturbance in the operation of the chemical action of the blood, the effect of which is called disease. To give names to these effects is the insanity of science.

The word Peter, or Petra, means "A rock." "Thou art Peter; on this rock will I build my church."

This statement, or word, represents the creative or formative principle defining the human organization. The twelve cell-salts of the body are stones, i. e., minerals, which in combination may be called a rock. These minerals, or rock, attract by chemical affinity the aerial elements, and by their union—chemical operation—the oil, albumen, fibrin, etc., which build up the human structure are formed and changed into bone and other tissue of the body, and thus build the beth, or church of God; the true church of God is the body.

The alchemists of old, whom we in our blindness have imagined were religious teachers, understood the real meaning of the statements: The human body is the temple of the living God; and again, the Holy Ghost dwelleth in you, and the kingdom of heaven is within you, and Our Father, who art in heaven. A temple and a church or beth (Beth-el) mean the same. Solomon's temple is a myth, an allegory or symbol of the human body, the temple of the living God. Originally it was the soul of man's temple, or the temple for the soul. Thus we can understand how the temple is built "without the sound of saw or hammer."

The seers, scientists, and alchemists of the early centuries of the Pisces, or water age, into which the sun and solar system entered about 2200* years ago, realized that, for about that period the inhabitants of the earth—souls in flesh—would be a lost race; that, while the earth was down deep in the Pisces air, dense and watery, the material thought would cognize from the individual concept, being so environed that the unity of being could not be realized.

A lack of the knowledge of the unity or completeness of being, or the perfection or completeness of the body, or temple of being, was symbolized by the allegory of the tem-

---

*The solar system entered Aquarius, an air sign, about the year 1900. Aquarius is "the Sign of the Sun, or Son, of Man in the heavens."

*Biochemistry*

ple or pyramid, the capstone rejected—or not yet placed in proper position.

The mineral salts—rock foundation of the human structure—have been rejected by the medical builders for 2000 years or more, but are now, as the earth swings into the air age, or the age of Spiritual Man, being recognized as the "Head of the Corner."

Thus we see why the beautiful name, Biochemistry, has shone forth from the slowly crystallizing carbon of dead and dying isms and pathies, and now glitters like a diamond in the crown of science.

Biochemistry is the "stone the builders rejected."

# FUNDAMENTAL PRINCIPLES

HERE are a few questions and answers that will enable the patient to grasp the fundamental principle of Biochemistry:—

First. What are the remedies you use? Answer—The inorganic salts, as found in healthy human blood.

Second. What is the meaning of Biochemistry? Answer—The chemistry of life.

Third. Where are the inorganic salts found? Answer—In all nature. In the food we eat; in the earth, rock, soil, and vegetable, and especially noticeable in mineral springs.

Fourth. Then why need we take them as medicine? Answer—You need not take them as medicine. No medicine, in the common use of the word, is or can be needed; they are taken as food, to supply a deficiency.

Fifth. Why does a deficiency occur, if the food we eat contains the mineral salts? Answer—Because the digestion and assimilation sometimes fail to set them free from the organic parts of the food, so that the absorbents can take in a sufficient quantity to keep the blood properly balanced; or some extra demand has been made upon the system—overwork, physical or mental, atmospheric or electric changes, etc.—which have too rapidly consumed the vitality of the body. It is then Biochemistry comes to the rescue. The inorganic vitalizing principles of food, having been set free by chemical process, or prepared direct from the mineral base, are given as a remedy and are taken in by the absorbents at once, not passing through the process of digestion at all, as they are ready for the blood when taken.

There is no such thing as disease, therefore, there cannot be any cure, as commonly understood. The symptoms called disease and named in Latin or Greek, so that the masses are awed and frightened by them, are not things or entities —are not something to be combatted, but are simply and

## Fundamental Principles

only the words, the dispatches, the language nature employs in calling for that which is lacking.

Do you see the difference between something and the lack of something?

"Yes," you say, "but people die from these words, dispatches, language, as you put it." Answer—Let us illustrate: A man goes without food for three days and nights, and has pains, fevers, headache, etc., but you know he is not "possessed of something," but lacks food. You also know he will die if the food is not supplied; but the words or dispatches calling for food will not kill him, but he will die because of a lack of food. You know this, and give him food; but, if you did not know it, you would proceed, according to the old pathology, to try to cure the pain, fever, or headache with some poison. "He asked for fish, and ye gave him a serpent."

# BIOCHEMISTRY AND THE BIO-CHEMIC PATHOLOGY

A SHADOW cannot be removed by chemicals, neither can disease be removed by poisons. There is nothing (no-thing) to be removed in either case; but there is a deficiency to be supplied. The shadow may be removed by supplying light to the space covered by the shadow.

So symptoms, called disease, disappear or cease to manifest when the food called for is furnished.

The human body is a receptacle for a storage battery, and will always run right while the chemicals are present in proper quantity and combination, as surely as an automobile will run when charged or supplied with the necessary ingredients to vibrate or cause motion.

There can be but one law of chemical operation in vegetable or animal organisms. When man understands and co-operates with that life chemistry, he will have solved the problem of physical existence.

When the arteries contain a sufficient quantity of the cell-salts, the aerial elements that form the organic portion of blood are drawn into them by chemical affinity or magnetic attraction, and precipitated or concentrated to the consistency that forms the substance known as blood.

The quality of blood depends entirely upon the chemical mineral base. If one or more of the inorganic salts are deficient in quantity, the blood will be deficient in vital or magnetic vibration and cell and tissue-building substance.

And so to supply the organism with the mineral principles that form the positive pole of blood is the natural law of cure.

Lymph and the lymphatic system is a part of the complex wonderful operation in the process of transmuting *the etheric substance*—aerial elements—into blood, flesh, and bone.

Biochemistry has been and is now being recognized by the most advanced thinkers the world has known.

# THE CHEMISTRY OF BLOOD AND TISSUE

THE word Biochemistry is formed from *bios*, the Greek for life and chemistry. Webster defines chemistry as that branch of science which treats of the composition of substances and the changes which they undergo. Therefore, Biochemistry, taken literally, means that branch of science which treats of the composition of living substances, both animal and vegetable, and of the process of their formation. But usage has given the word a somewhat different signification, and the following is a more accurate definition: That branch of science which treats of the composition of the bodies of animals and vegetables, the processes by which the various fluids and tissues are formed, the nature and cause of the abnormal conditions called disease, and the restoration of health by supplying to the body the deficient cell-salt.

The chemical composition of tissue and the various fluids have long been known, but, until Biochemistry was introduced, no practical use had been made of this knowledge in the treatment of the sick. The so-called science of medicine has no claim to the name, science. We refer to the old system that treats disease as an entity—a something, or at least caused by a something instead of a deficiency, which all will admit to be a lack of something. It is useless for those who adhere to the practice of the drug system to try to defend it. We have the testimony of many of their most noted professors and authors, that their "system of practice is responsible for more deaths than war, pestilence, and famine combined."

We realize that modern surgery is an exact science. Like watch making or house building, it is purely mechanical. In anatomical exactness, and in instruments of precision, the advances in surgery during the past fifty years have been marvelous.

## The Chemistry and Wonders of the Human Body

While the diagnosis of disease by surgeons is many times at fault, sometimes fatally so, yet their mechanical operations are beyond criticism.

Homeopaths b u i l t better than they knew. In preparing their high potencies, they eliminated the poison contained in such drugs as aconitum and belladonna, and left only the inorganic cell-salts which supply deficiencies when correctly selected. But this subject is dealt with in the book, "The Biochemic System of Medicine," which every student of course reads.

Biochemistry is science, not experimentalism. There is no more of mystery and miracle about it than about all natural laws. The food and drink taken into the stomach and the air breathed into the lungs furnish all the materials of which the body is composed. By the juices of the stomach, pancreas and liver, the food is dissolved, and the cell-salts are taken up by the absorbents and carried to the lungs, where they unite with the aerial elements and make blood.

When we realize that there is as much matter thrown out of the body in twenty-four hours as is taken into it, we see that flesh is not formed from the food we eat.

Oil taken into the stomach can not possibly reach the tissue as oil, simply because it passes through a metamorphosis from the action of the gastric juice, bile and pancreatin.

The blood supplies the material necessary for forming every tissue and fluid in the body, and for carrying forward every process in the operation or materialization of the human form.

An analysis of the blood shows that it contains organic and inorganic matter. The organic constituents are sugar, fats and albuminous substances. The inorganic constituents are water and certain minerals commonly called cell-salts or tissue builders. Of a living human body water constitutes over seven-tenths, and cell-salts about one-twentieth, organic matter the remainder.

The writer was among the first scientists in the world to advance the theory (now a demonstrated scientific truth) that the organic portion of all vegetable and animal matter,

## The Chemistry and Wonders of the Human Body

the oil, albumen, fibrin, etc., is *formed by a combination of aerial elements* which make up what is known by the general term atmosphere. Food taken into the system is not changed to flesh and bone, but acts as a negative pole, as explained in the following extract from one of Dr. Carey's lectures:

"The commonly-accepted idea that vegetation is a product of soil, that it absorbs from the earth the material that builds the structure of the plant, and that animal tissue is built up by a metamorphosis of this vegetable substance into flesh and bone, has been proven erroneous.

"Chemistry and the spectroscope prove that *vegetable* and *animal tissue is precipitated air.*

"It is well known by chemists that all manner of fruits, grains and vegetables are produced directly from the elements in the air, and not from the soil. The earth, of course, serves as a negative pole, and furnishes the mineral salts of lime, magnesium, iron, potassium, sodium, and silica, which act as carriers of water, oil, fibrin, sugar, etc., and thus build up the plant. But the oil, sugar, albumen, etc., are formed by a precipitation of principles in the air, and not from the soil. This is a fact abundantly proven. M. Berthelot, a scientist of France, Tesla, the Austrian wizard, and our own Edison have long held that food could be produced artificially by a synthetic process from its elements. Some six or seven extracts, as well as coloring material, are now being manufactured in this manner. Madder is now made almost exclusively by this process.

Long ago I advanced the theory that animal tissue is formed from the air inhaled, and not from food. The food, of course, serves a purpose; it acts as the negative pole, as does the earth to plant and vegetable life, furnishes the inorganic salts, the workers that carry on the chemistry of life, and sets free magnetism, heat, and electric forces by disintegration and fermentation of the organic portions of the food. But the air in passing through the various avenues and complex structure of the wonderful human organism, changes, combines with the mineral salts and solidifies, until it is finally deposited as flesh and bone."

# CHEMISTRY, ELECTRICITY, OR SOLAR ENERGY

THAT which is generally called electricity, is only the effect, or manifestation of energy.

Where, or what, is the source of this power or energy?

The sun-solar energy.

In what manner does this energy operate to produce the phenomena of light and heat?

By its operation or chemical action on so-called matter, i. e., the aerial elements. Different rates of vibration produce different manifestations—heat, cold, light, so-called electrical effect, sound, color, animal or plant growth, blood, etc. Air breathed into the arteries (air carriers) unites with the mineral base of the blood, the inorganic salts of food, and is thus precipitated, condensed, and chemically changed into blood by the same law (Infinite intelligence) that changes these elements into vegetables, fruits, nuts, flowers, grass, etc.

The organic portion of food—oil, albumen, fibrine, etc.—is consumed, chemically burned up in the stomach and intestinal tract, to set free its stored-up energy for motive power to run the human laboratory, or machine, so the process of inhaling air,—raw material for blood—may go on. By this combustion, the mineral, or cell-salts, of iron, lime, potassium, magnesium, sodium and silica, are set free, and enter the blood-vessels by transmosis, and form the negative pole of the chemical formulae, called the blood. This blood is the product of energy operating upon matter (which is energy in concrete form), and proves the ancient statement true that "God made of one blood all nations that dwell on the earth." This is LITERALLY TRUE, for there is only *one substance* from which to make anything.

## Chemistry, Electricity, or Solar Energy

Then are we to understand that solar energy and electricity are one and the same? Yes. There is but one energy, one source of power in the universe. It is the one and only Dynamis. This energy is neither electricity, heat, light, darkness, or cold, but produces all these and all other phenomena by its word, will, operation or vibration. There is absolutely no proof that the sun is hot, but there is abundant proof that it is a mighty center, or dynamo, of energy, force or power, constantly radiating its waves of energy throughout the solar system, and the action of this force, or friction, on the aerial envelope of the earth causes heat in different degrees, according to the divine, creative will.

Two clear days in August, the temperature one day is 95 degrees, the next 75 degrees.

Cause: Different rates of vibration in the waves of energy from the great solar center dynamo—the sun—directed by Infinite intelligence.

Does it not seem very foolish to complain continually about the weather?

How does electricity, or energy, pass along a wire?

It does not pass at all. Electricity is a myth, so far as being a substance is concerned. So-called electricity is an effect. That which causes the effect is stationary, "the same yesterday, today, and forever.' It is substance (the body of God) in operation. Etheric atoms are everywhere; therefore the universe is solid. Disturb this body of atoms at any given point and ALL is disturbed. This disturbance, jar, motion, quiver, or vibration, records itself when it impinges upon a resistant point sensible or delicate enough to cognize or respond to its operation.

Place your hand upon a piece of timber or iron bar, and have some one strike the other end a blow with a hammer, and you will instantly feel a jar or vibration, but nothing—no substance—passed from one end to the other. The molecules, or particles that compose the wood or metal, vibrate each other, and thus produce the same motion at the opposite end. This explains the science of telegraphy, the telephone, etc. The vibration set up or started at one end of a wire set every particle, molecule, or atom of the wire in vibration; thus characters or sounds may be recorded.

## The Chemistry and Wonders of the Human Body

Flesh, bone, blood, nerves and all fluids of the body, are composed of cells, molecules, or atoms, formed from the same universal etheric substance that composes the molecules that make the wire, and the rate of vibration set in motion by the dynamo is transferred to the human organism, and a jar or shock is the result.

The reason that glass or rubber is a non-conductor of this jar, shock or motion, quiver or vibration, is simply because the particles that make up these substances are not arranged in a manner to move or oscillate in their place, or at least, but very slightly.

Wireless telegraphy is explained upon the same principle. There is a universal substance, everywhere, in molecules, between which is universal energy (the body and spirit of the universe), and vibrations can be conducted through so-called air, and recorded, provided a receiver corresponding with the sender, is prepared.

Is wireless telegraphy the last step in the science of communication? No. The brain of man (and woman) is both a transmitter and receiver, and when the race awakens, it will find that the brain is the only instrument needed, and that the medium for transmission of thought is everywhere present.

# CELLULAR PATHOLOGY

ALL diseases that are curable are cured in a natural manner through the circulation; the constituent parts of the human organism, which are carried by the blood vessels and transude through the walls of these "branches of the tree of life" into the surrounding tissue, restore normal conditions when the blood contains the proper amount of water, sodium, ferrum, potassium, calcium, magnesia, and silica.

Deficiencies in the cell-salts produce pains, fever, spasms, or some other cry of distress. These so-called symptoms are words, or dispatches, calling for what is needed. And when the call is for the phosphate of potassium to supply nerve cells, shall we give morphine? "He asked for bread, and ye gave him a stone."

When the call is for the phosphate of iron, in order that more oxygen may be conducted through the organism, and thereby increase vitality, shall we give alcohol? "He asked for fish and ye gave him a serpent."

It will be observed that there is nothing miraculous about the biochemic procedure—it is simply natural law.

The constituents of our bodies, planned by Infinite Intelligence, keep all parts of its wondrous mechanism in harmonious co-ordination when present in proper magnitude and amount. Harmony cannot be obtained when deficiencies exist by introducing a poison into the system.

The symptoms may be changed to those that manifest differently, but the patient is not cured.

Calomel does not cure; it simply sets up a diarrhoea in place of constipation. Opium does not cure; it sets up paralysis of nerve centres in place of neuralgia (Greek and Latin for nerve pain).

We do not claim magical curative properties for the biochemic *materia medica*. We only point out the law of the chemistry of life.

## The Chemistry and Wonders of the Human Body

Is there any system of teaching or practice before the world today that can be said to present the law of cure as full-rounded and many-sided, clear, explicit, without evasion or ambiguity, as does the science of Biochemistry?

Let the sick bear in mind that there is but one way to be restored to health, and that is the *natural* way: through the blood by *supplying deficiencies*. It will require just as much time to cure as nature requires, working in a *natural* way. The food or workers called for *must* be supplied; calomel, aconite, belladonna, salicylic acid, opium, etc. (we do not refer to the homeopathic triturations of these drugs) are not constituent parts of blood, are not found in the human organism naturally, and, when taken into the system, set up their own vibrations or action, in place of the condition naturally produced by a deficiency in the component parts of the organism, and are worse than the disease for which they are given. When a twig is broken from a branch, we know a new one will grow again to the same size, if water is supplied to the soil and conditions favorable to its growth are furnished; we do not expect to supply a new growth by legerdemain, or some short cut—say by putting an "active poison" about the roots of the tree—or injecting beneath its bark a nameless lymph wherein sport the festive bacilli and all-pervading microbes.

We realize the branch must be restored in a natural manner by the constituent parts of the tree, operating or circulating through the physiology of the tree, and thus carrying on the process of growth.

# THE VIEWS OF SCIENTISTS

PROFESSOR VIRCHOW, in his lecture on "Cellular Pathology," says (see lecture 14) "the cells of the organism are not fed, they feed themselves. The absorption of matter into the interior of the cells is an act of the cells themselves."

Alfred Binet, a noted French scientist, says, in his work, "The Psychic Life of Micro-organism;" "The micro-organisms do not nourish themselves indiscriminately, nor try to feed blindly upon every substance that chance may throw in their way. The microscopic cellule in some manner knows how to choose and distinguish alimentary substances from particles of sand."

So I am led to believe that the cells are intelligent organisms and can choose their nourishment. This being the case, how foolish, if not criminal, to place only a poisonous agent within their reach.

As the researches of Binet, the French scientist, show that micro-organisms—infusoria—select their own food from the material at hand, so does the Pomeranian scientist, the great Virchow, clearly demonstrate that the cells that build the human form divine also select their nourishment from material within reach, and that nothing foreign to their constituent parts can be forced upon them—except to produce injury or death. Professor Virchow's researches demonstrate the fact that abnormal cells are caused by a lack of the chemical constituents that are required to produce normal cells.

The renowned Dr. Schuessler says: "The inorganic substances in the blood and tissue are sufficient to heal all diseases that are curable at all. The question whether this or that disease is or is not dependent on the existence of germs, fungi, or baccilli is of no importance in biochemic treatment. If the remedies are used according to the symptoms, the desired end, that of curing disease, will be gained in the shortest way. Long-standing chronic diseases, which have been brought about by overdosing or use of poisonous drugs, quinine, mercury, morphine, alcohol, etc., may be cured by minute doses of cell-salts."

Professor Liebig, the world-wide authority in chemistry, says: "It happens that a tissue in disease reaches such a degree of density, becomes so clogged, that the salt solution of the blood cannot enter to feed and nourish; but if for therapeutic purposes, a solution of salt be so triturated and given so diluted that all its molecules are set free, it is presumable that no hindrance will be in the way of these molecules to enter the abnormally condensed part of tissue."

The body is made up of cells. Different kinds of cells build up the different tissues of the body. The difference in the cells is largely due to the different mineral salts that enter into their composition. If we burn the body, or any tissue of it, we obtain the ashes. These are the mineral or inorganic constituents of the body, the salts of iron, lime, magnesium, etc. They are the *tissue builders*, and both the structure and vitality of the body depend upon their proper quantity and distribution in every cell.

Professor Huxley said: "Those who are conversant with the present state of biology will hardly hesitate to admit that the conception of life of one of the higher animals as the summation of the lives of a *cell aggregate*, brought into harmonious relation and action by a co-ordinative machinery formed by some of these cells, constitutes a permanent acquisition to physiological science. I believe it will, in a short time, become possible to introduce into the human organism a molecular substance that will by the law of chemical affinity find its way to the particular group of cells or nerve plexus that may be in need of it."

We know that the cell- salts, or mineral workers, in the blood are infinitesimally sub-divided in the food we eat. Nature works everywhere with immense numbers of infinitely small atoms, which can only be perceived by our dull organs of sense when presented to them in finite masses. The smallest image our eye can recognize is produced by billions of waves of light. A granule of salt which we can

## The Views of Scientists

scarcely taste contains millions and millions of groups of atoms which no human eye will ever discern. A search for the ultimate atom will surely end in complete recognition of the operation of Wisdom, or Omnipotent Life.

One quart of milk is found by analysis to contain about the six-millionth of a grain of iron; a child fed on milk receives each time one milligram of iron in a half-pint of milk, which is only the fourth part of the above minute fraction of one part of a grain of iron. But this infinitesimal amount supplies the iron molecules needed to carry a full supply of oxygen to all parts of the organism.

The proportion of fluorine in the human organism is still less than that of iron.

Professor Lieberg says: "Hydrochloric acid, diluted with one thousand parts of water, readily dissolves the fibrin of meat and the gluten of cereals, and this solvent is *decreased* when the acid solution is made stronger." One red-blood corpuscle does not exceed the one hundred and twenty-millionth of a cubic inch. There are over three millions such cells in one droplet of blood, and these cells carry iron and other mineral workers. How necessary, then, to administer these salts in minute molecular form!

# THE BIOCHEMIC PATHOLOGY OF EXUDATIONS, SWELLINGS, ETC.

IT is very important for the student of Biochemistry to fully understand the real cause of exudations, swellings, inflammation, eruptions and all accumulations of all so-called morbid matter which accompany disease.

Heteroplasm is defined by Virchow as a substance foreign to the normal constituent parts of the human organism. It must be that the normal constituent parts of the body have become abnormal for some reason, as the change occurs in the organism of man and not outside.

The fluids of the body, containing oil, fibrin and other albuminous substances, could not so combine as to render them non-functional, or to form vitiated compounds, causing exudations, eruptions, etc., if the vitalizers, the workers, called inorganic cell-salts, were present in proper quantity. There is a small amount of muriatic acid in gastric fluids, and a deficiency of the same causes indigestion and possibly catarrh of the stomach, which means exudation of certain organic matter from the blood; but this abnormal state would not exist unless there first arises a deficiency in some one or more of the inorganic or mineral salts. Muriatic acid is formed by the union of certain dissimilar substances, which conjoin and form new compounds. This is Biochemistry—that is, life chemistry, the real definition of the word.

Any breaking up of the chain of continuity of the cell-salts, of course, disturbs this process of forming new organic material to replace that cast off as worn out and useless.

Let us take a condition called acidity, or an excess of acid. It is well known that either acid or alkali alone may be injurious to man and seriously interfere with the process of life, while a proper combination of the same forms natural fluids of the body.

*The Biochemic Pathology of Exudations*

There is not, properly speaking, an excess of acid, but a deficiency in the alkaline cell-salts.

A deficiency of sodium phosphate causes a breaking up of the basis of certain parts of the blood plasma, which causes, or allows, the acid to stand alone and thus produce a disturbance. Of course, this is due entirely to a lack of the alkaline salts—a proper balance of them—and not to an excess of acid. When the true cause of eruptions, swellings, exudations, etc., is fully understood, the names now used to designate the supposed difference will not be used.

Instead of treating a certain disease (which simply means not at ease), the physician will learn the telegraphy of the intricate, complex, marvelous human machine and know what is asked for in the words now, or heretofore, supposed to simply be pain or exudations.

# CHEMICAL OPERATION

A CHILD may touch a button that will start a complex machine to operating, and yet not understand the science of physics or the mechanism of the machine.

So many systems of healing may be the means of starting the workmen in the system that have become dormant into action. Massage, bathing, electricity, magnetic healing, suggestion, absent treatments, concentration, affirmation, prayer, all these and many more that might be mentioned, can and often do *start* forces that have become dormant because some link in the chemical chain of inorganic molecules has been misplaced or thrown out of gear, *but when these chemicals (man's body is a chemical formula, remember) are deficient* in the blood, you can no more supply them by any of these modes of operation than you can cure hunger by them. These methods are all good in their time and place to start dormant energies, but none of them will supply deficiencies—viz., cure hunger.

So Biochemistry furnishes the key to all cures made by the old or allopathic, the homoeopathic or electic schools, or by medical springs or healing through the operation of mind.

There are some who heal by thought transference, others must come in contact with the patient. In either case, I hold the process is orderly and within the domain of law.

This science is in perfect harmony with the Chemistry of Life operating in each human organism, and cannot antagonize any phase of higher thought. Mind, or mental cures, Christian or Divine science, suggestive therapeutics or magnetic healing, must all operate according to Divine law (Life Chemistry), or not at all. The operation of wisdom has many names, but the *chemical process* is one.

# MAN'S DIVINE ESTATE

### A Prophecy of the Age of Alchemy

BIOLOGISTS and physiologists have searched long and patiently for a solution of the mystery of the differentiation of material forms.

No ordinary test can detect any difference in the ovum of fish, reptile, animal, bird, or man. The same mineral salts, the same kind of oil, albumen, fibrin, or sugar, or carbon is found, not only in the egg or germ of all forms of life, but in the substance or tissue of the bodies of all the varied expressions of materiality.

The answer to this "Riddle of the Sphinx" is found where Bio, or Life Chemistry merges into alchemy, over the door of which is written, "It is finished"—"Let there be light."

Professor Loeb says: "The ultimate source of living matter is chemical." To the Biochemist the above is a truism. There is no such thing as inert, or dead matter. All is life.

The base of all manifestation is mineral. Out of the dust, ashes, or minerals of the earth physical man is made.

The twelve mineral salts are the basis of every visible form, animal or vegetable. No two different forms have exactly the same combination of the minerals, but all have the same minerals.

These minerals, inorganic salts, are the twelve gates of precious stones described by John in his vision. When the Divine Word speaks the mineral atoms, or molecules, of its body into a certain formula or combination, a germ or egg, which is the basis or nucleus of the form to be manifested, materializes. This little plexus of intelligent atoms then commences to attract to its centre by the law of chemical affinity, which is only another way of saying God in action, other atoms known as oxygen, hydrogen, nitrogen, etc., and thus materializes them, until the building is completed according to the plan of the architect or designer. Thus the

## The Chemistry and Wonders of the Human Body

Word, operating through chemistry, is the Alpha and Omega.

There would be no eagle, fish, horse, or man without the Word, Divine Wisdom, and there would certainly be no Word, if there was no substance to obey the Word, and likewise there would be no substance, if there was no law of chemical affinity, or action and reaction, whereby the operation of materialization and dematerialization may be carried on.

It will be demonstrated in the near future, that so-called nitrogen is mineral in solution, or ultimate potency, which explains the reason why nitrogen enriches the soil.

The Atmospheric Product Company at Niagara Falls, whose promoters expect to extract and condense nitrogen from the aerial elements by electrical process, are the forerunners of machines that will manufacture our food and clothing direct from the air, and also produce heat or cold as needed by different rates of vibration of the substance, body of God, everywhere present.

Neither light nor heat comes from the sun, for they are not entities that can come or go. They are effects or results. The sun is surely a great dynamo, or vibrating centre of Divine Energy, which by its thought moves the atoms of our earth-envelope, and by vibration or motion chemically causes light and heat in different degrees, according to its good will and pleasure.

By chemistry, the court of last resort, will man come into his divine estate. He will then place the "Poles of Being," and produce vegetable or animal forms at will.

Thus the prophecy of man's dominion will be fulfilled, for he will have attained knowledge which will enable him to manufacture a psychoplasm (if I may be permitted to coin a word), from which he can bring forth all manner of vegetable or animal life.

> Let man stand upright and splendid,
>   Let woman look up from the sod—
> For the days of our bondage are ended,
>   And we are at one with God.

# INTUITION, OR THE INFINITE VIBRATION

INTUITION is information direct from the source of all knowledge, vibrating the brain cells and nerve centres of the human organism—the temple, or instrument, of the living God—at different rates, or tones, according to the chemical combination of physical atoms composing the organism. An alligator, a horse, a monkey, or a man, being organized each on a different key, receives and expresses the infinite word according to its note, molecular arrangement, or chemical formula.

Wireless telegraphy is demonstrating the underlying principle of what the world has named intuition.

Mental, or absent healing, is scientifically explained in the explanation of wireless telegraphy. The same substance—air, or ether—fills the so-called space in which we exist. We (our bodies) are strung on this attenuated substance like spools on a string; it extends through us—we are permeated with it as water permeates a sponge.

When the brain cells of the mental healer are acted upon by the word or thought of the healer, they vibrate, jar, or oscillate at the rate that causes an arrangement of cells that manifests or materializes the bodily functions on the plane of health. This rate of vibration, started from the sender, will produce the same rate, or jar, in the brain cells of any one attuned to the note, providing they recognize the operation, for it is only through such consciousness that we become a receiver.

Thus we realize the truth of the statement, "Thy faith hath made thee whole."

The sender and receiver in the Marconi system must be in the same key—that is, adjusted to sense the same jar, or vibration. When the brain of the healer and patient are in unison, through conscious understanding or agreement, cures can always be effected, if the chemical constituents—molecules—are present in the organism of the patient, though dormant, negative, or out of harmonious co-ordination, by the proper jar, or thought vibration, of the healer or sender.

But, if the blood of the patient is really deficient in some of the mineral salts, or cell-salts, the phosphates, sulphates, and chlorides of iron, lime, potassium, and other inorganic substances which compose the material organism, the cure cannot take place unless the jar, or vibration, of health started by the sender so oscillates, or jars, the fluids of digestion and assimilation that these lacking elements are set free from the food and water taken by the receiver (patient) and thus supply the deficiency. It is these cases that baffle mental or divine healing.

Biochemistry fills the gap, offers the solution of the problem by preparing the cell-salts and proceeding directly to the work of doing that which is absolutely necessary to be done, namely, supplying *the necessary chemical molecules.*

The body is a storage battery, and must be supplied with the necessary chemicals, or it will not run. When this can be done by right thinking, well and good; but, when a deficiency does occur, why not supply it direct by a biochemic procedure? Mental science is an incomplete science without Biochemistry.

# THE SO-CALLED ELEMENTS

SO far as science has been able to weigh, measure, or in any manner cognize them, there seems to be about seventy-two elements in nature, viz., in the earth, water, and air, a combination of which makes manifest all we see or recognize in the material world.

We do not print the following as an ultimatum, for certain changes in the aerial elements, as well as in the cellular structure of the human brain may, and doubtless will, enable man to penetrate deeper and deeper into the Holy of Holies of Omnipresent Life.

We do not believe there is any such thing as an element different from the universal substance in its last analysis. We believe that so-called "elements" are different rates of motion, or vibration of *one substance*.

> Behind all chemical phenomena,
> "Standeth God within the shadow,
> Keeping watch above His own."

The names and abbreviations of the so-called principles or elements are as follows:

| | | | |
|---|---|---|---|
| Aluminum | Al | Fluorine | F |
| Antimony (Stibium) | Sb | Glucinum | G |
| Argon | Ar | Gold (Aurum) | Au |
| Arsenic | As | Hydrogen | H |
| Barium | Ba | Helium | Hm |
| Bismuth | Bi | Indium | In |
| Boron | B | Iodine | I |
| Bromine | Br | Iridum | Ir |
| Cadmium | Cd | Iron (Ferrum) | Fe |
| Caesium | Cs | Lanthanum | La |
| Calcium | Ca | Lead (Plumbum) | Pl |
| Carbon | C | Lithium | Li |
| Cerium | Ce | Magnesium | Mg |
| Chlorine | Cl | Manganese | Mn |
| Chromium | Cr | Mercury (Hydrargyrum) | Hg |
| Cobalt | Co | Molybdenum | Mo |
| Copper | Cu | Nickel | Ni |
| Didymium | Di | Niobium | Nf |
| Erbium | Er | Nitrogen | N |

## The Chemistry and Wonders of the Human Body

| | |
|---|---|
| Osmium ........................Os | Sulphur ..........................S |
| Oxygen ..........................O | Tantalum ......................Ta |
| Palladium ....................Pd | Tellurium ......................Te |
| Phosphorus ....................P | Thallium ........................Ti |
| Platinum ......................Pt | Thorium ........................Th |
| Potassium (Kalium)..........K | Tin (Strannum)................Sr |
| Radium ........................Rm | Titanium ........................Ti |
| Rhodium ......................Ro | Tungsten (Wolfram)..........W |
| Rubidum ......................Rb | Uranium ..........................U |
| Selenium ......................Se | Vanadium ........................V |
| Silicon (Silica)................Si | Yttrium ..........................Y |
| Silver (Argentum)............Ag | Zinc ..............................Zn |
| Sodium (Natrum)..............Na | Zirconium ......................Zr |
| Strontium ......................Sr | |

Ten or twelve more might be added, but their atomic weight and affinities are so uncertain that we have omitted them for the present.

# ESOTERIC CHEMISTRY

IN this strenuous age of reconstruction, while God's creative compounds are forming a new race in the morning of a new age, all who desire physical regeneration should strive by every means within their reach to build new tissue, nerve fluids and brain cells, thus literally making "new bottles for the new wine." For be it known to all men that the word "wine" as used in Scripture, means blood when used in connection with man. It also means the sap of trees and juice of vegetables or fruit.

The parable of turning water into wine at the marriage at Cana in Galilee is a literal statement of a process taking place every heart beat in the human organism.

Galilee means a circle of water or fluid—the circulatory system. Cana means a dividing place—the lungs. In the Greek, "A place of reeds," or cells of lungs that vibrate sound.

Biochemists have shown that food does not form blood, but simply furnishes the mineral base by setting free the inorganic or cell-salts contained in all food stuff. The organic part, oil, fibrin, albumen, etc., contained in food is burned or digested in the stomach and intestinal tract to furnish motive power to operate the human machine and draw air into the lungs, thence into the arteries, *i. e.,* the air carriers.

Therefore, it is clearly proven that air (spirit) unites with the minerals and forms blood, proving that the oil, albumen, etc., found in the blood, is created every breath at the "marriage in Cana of Galilee."

Air was called water or the pure sea, viz.: Virgin Mar-y. So we see how water is changed into wine—blood—every moment.

In the new age, we will need perfect bodies to correspond with the higher vibration, or motion of the new blood, for "old bottles (bodies) cannot contain the new wine."

## The Chemistry and Wonders of the Human Body

Another allegorical statement typifying the same truth reads, "And I saw a new Heaven and a new Earth," *i. e.*, new mind and new body.

Biochemistry may well say with Walt Whitman: "To the sick lying on their backs I bring help, and to the strong, upright man I bring more needed help." To be grouchy, cross, irritable, despondent, or easily discouraged, is prima facie evidence that the fluids of the stomach, liver and brain are not vibrating at the normal rate, the rate that results in equilibrium or health. Health cannot be qualified, *i. e.*, poor health or good health. There must be either health or dis-health; ease or disease. We do not say poor ease or good ease. We say ease or dis-ease, viz., not at ease.

A sufficient amount of the cell-salts of the body, properly combined and taken as food—not simply to cure some ache, pain or exudation—forms blood that materializes in healthy fluids, flesh and bone tissue.

The microscope increases the rate of motion of the cells of the retina and we see things that were occulted to the natural rate of the vibration of sight cells. Increase the rate of the activity of brain cells by supplying more of the dynamic molecules of the blood, known as mineral or cell-salts of lime, potash, sodium, iron, magnesium, silica, and we see mentally, truths that we could not sense at lower or natural rates of motion, although the lower rate may manifest ordinary health.

Natural man, or natural things, must be raised from the level of nature to super-natural, in order to realize new concepts that lie, waiting for recognition, above the solar-plexus, that is, above the animal or natural man.

The positive pole of Being must be "lifted up" from the Kingdom of Earth, animal desire below the solar-plexus, to the pineal gland that connects the cerebellum, the temple of the Spiritual Ego, with the optic thalamus,, the third eye.

By this regenerative process millions of dormant cells of the brain are resurrected and set in operation, and then man no longer "sees through a glass darkly," but with the Eye of Spiritual Understanding.

Biochemistry is the sign-board pointing to the open country, to the hills and green fields of health, and the truth that

## Esoteric Chemistry

shall set the seeking Ego free from poverty and disease. The power to heal resides in each living cell of tissue, because the cells are certain manifestations of life, or spirit. They indicate a certain step in divine manifestation. This power of the cell is conferred upon the whole organism; when an injury occurs to the flesh, the nerves convey a sensation to the brain that immediately produces a flow from the nutrient arteries, and these carry the blood, freighted with life material, and deposit it in the wound, and thus rebuild.

How absurd, then, to think that a plaster or salve, or liniment can heal. It is natural to get well and unnatural to remain sick or injured, or die. These facts are admitted and taught by the leading physiologists and medical men of the present day, but they are now coming into increasing prominence. "The healing power of Nature" or *vis medicatrix naturae*, is an old phrase in medical science.

Professor W. J. Youmans, of the *Popular Science Monthly*, says: "All who have watched the progress of the healing art in recent times, will note that among the most enlightened practitioners there has been a steadily diminishing confidence in medication and an increasing reliance upon the sanitary influence of Nature."

Professor B. W. Richardson, M. D., one of the most eminent of British physicians, in an address before the Sanitary Institute, in 1878, used this strong language:

"The science of prevention must take all the world with it. It becomes a political and social, as well as a medical study, appealing to all minds. It models itself into household truths, and commingles with the moral and religious elements of life. I need not say that the pathies must go. The pathies of all kinds are dead as door-nails, and wait only to be interred decently in a common grave. In time, the word 'cure' will go altogether. With the progress of sanitary science, caused by indulgences, or overwork, it will be removed by the effect of moral and spiritual influences and a knowledge of causes."

# THE FALLACY OF THE GERM THEORY OF DISEASE

THE true physician says, with Emerson, "I will proclaim what I prove to be true today, though it contradict what I have advocated all my life."

Do we try to use reason in figuring out any problem? Long ages of working for self-preservation have taught mankind that when a certain sensation is felt at the pit of the stomach it is caused by hunger, and that, if food is taken, the symptom will disappear.

But the hungry one does not once think of a germ or microbe causing that gnawing sensation. He realizes, intuitively, that the human laboratory is sending out an S.O.S. to let the owner know that it needs fuel for its furnace, and that, given the fuel, combustion will at once begin and heat and energy will be furnished to run the human dynamo. At the same time, in this process of generating heat and energy, the mineral particles which are contained in the food will be set free, taken up by the absorbents and carried into the blood, the river of life, and distributed to all parts of the body.

As this wonderful river flows through, each tiny individual cell reaches out and actually selects from the stream, the inorganic cell-salt that it needs to re-build that portion of itself which has been destroyed in the continuous process of tearing down and building up.

If each cell is furnished with material to replace that which is torn down, then there is no interruption of the orderly process of nature and consequently there is perfect equilibrium.

When we repair a house, or any article of use, we select the same kind of material that was employed in its construction. Why should not this rule hold good when applied to the "Temple of God"—the human body? When we stop

## Theory of Disease

to really think about it, does it not seem unreasonable and absurd to give it poisons and drugs of any kind?

No actual progress has been made in the definite cure of any disease. In surgery, however, which is mechanical in its operation, marvelous progress has been made. Take such a common disease as pneumonia, for instance; one is never sure whether the patient will live or die. If the symptoms do not cease, it is evident that the body has not been supplied with the proper material, else it would not be sending out signals, calling attention to itself.

If the little cells of the body, in their reaching out to grasp from the blood that which they need, do not find it, they do not take anything. What is the result when a person cannot get food? We do not say he has "caught" hunger—that a certain germ has caused it, or that he can give it to another, or that another can "catch" it from him. But when the cells cannot get what they need, we say that the person has "caught" some kind of disease—some microbe has entered the laboratory of the human body. Does it seem reasonable?

What really occurs is this: The cells break down and disintegrate, because there is nothing, or not enough material given, to supply or replace the waste. As the functions are disturbed, this waste matter is not carried out of the body, but remains, clogging and poisoning the system. The little drain tiles which carry off the excess moisture and the poisons of the body are clogged and the surplus heat and moisture cannot escape; then fever, so-called, results. If these little vents of the human body are not soon restored to their normal function, the patient dies. Give a stove too much fuel and close all the drafts, and the fire will go out. A person who eats more than the body requires, keeps the intestinal tract filled with fermenting food (just as a stove or furnace may be filled with clinkers), for there is not enough gastric juice to break it up and digest it. Auto-intoxication results and the intestinal tract is flooded with poison. Thus it is that such a person is always among the first to "catch" any so-called disease.

We give the engines that run our machinery a great deal more intelligent care than we give the dynamo that runs the

## The Chemistry and Wonders of the Human Body

physical machinery. A good machinist spends much time over his engine. He uses reason and common sense and knows that if his engine is clogged with oil or dirt it will not run easily and smoothly and he cannot expect to get maximum power from it. Why should physical machinery differ in this respect from artificial mechanism? Instead of using reason and giving the physical furnace the right proportion and quantity of fuel—what do we do? We eat, because it tastes good, a greater amount than we need, often suffering extreme discomfort.

When an article of food spoils, or a dead body starts to decompose, we do not say that a germ or microbe caused it—we know that it was because of atmospheric conditions; but we do realize that flies and maggots are formed from the decaying mass, and, if left alone, will consume it and die. The flies and maggots did not come until there was dead (so-called) matter present, for that is nature's method of removing it.

There are many kinds of material in the human body, and each is used for a definite purpose. For instance, there is much fibrin, which is a constituent of the tissue of the body. This needs the inorganic cell-salt, chloride of potassium (Kali Mur), to combine with it and make it functional. If the system is deficient in this cell-salt, the fibrin is not able to do its work—becomes non-functional—and, if not carried off by any of the means which Mother Nature employs, it clogs the mechanism. Then we have a cough, or pneumonia sets in, or perchance we "catch" typhoid. The same thing causes it all, namely, a deficiency in chloride of potassium. Microbes are found—yes—but they are the product of decaying matter.

Away off in some secluded mountain home, one of the members of a family is taken with typhoid and dies. None of the others have it. If *caused* by a germ, what prevented the entire family from "catching" it?

Of all negative conditions the race is subject to, fear is the greatest. We are born cowards. Our mothers feared for us, and their own lives, before we were born. We came into the world, with a wail of fear. All who had anything to do with us feared something would happen to us. They

## Theory of Disease

were afraid we would catch cold, or the measles, or whooping cough, or die with summer complaint. Somebody feared all the time that we would get burned, or fall out of the cradle, or into the well. We were afraid of our parents or teachers; that lovers would forsake us, or we would not have our lessons, or that we would catch cold, or that it would rain, so we could not go to the picnic. Still later, we were afraid of failure in business, or that our house or store would catch fire; afraid on sea and land; of fire and water, lightning, hail, wind, and all things visible and invisible, and yet we wonder why we are sick, and why humanity seems to be a failure.

Men and women are rising out of these conditions, and have begun to think; so look out for them. They will make a new earth.

Ignorance and fear are the causes of the spread of contagion, or, the shadows called "disease."

Some noted person, in Russia say, takes on a negative condition, is dosed with poisons, and dies. The so-called Scientists give it a name—"influenza," or "la grippe." The doctors, the people, and the press, describe the symptoms, and forthwith others "take it," then it is telegraphed to Berlin, Paris, London, New York and San Francisco, that la grippe, or the Russian influenza (genuine Russian) is raging across the Atlantic, and is very fatal; then a General, or Senator, or Governor, on this side of the ocean, dies, and a minute description of the symptoms and the treatment is given. Doctors polish up their microscopes and try to find, as Bill Nye said: "Some hitherto unidentified microbe." They search for germs and microbes in the sputa of the patient, and, of course, find them. All decaying organic matter, oil, albumen, etc., swarms with micro-organisms. The organized matter has disintegrated and created these minute individual entities. Their coming into being was a step in the process of the metamorphoses necessary to resolve organic matter back to the higher potency of spirit from whence it sprang. These microbes are found in an exudation from any orifice of the body. If you have only "caught" an ordinary cold, these Scientists never think of looking for the germ that caused you to "catch it;" but if you have it bad enough to name it in French, the germ becomes at once illustrious, and sees its name in the papers. So the people agree that the disease is caused by something so small that it requires a microscope to see it, but still can cross oceans and continents against the gale, and go straight down the throats of the helpless, terror-stricken people. The people are warned to "look out for la grippe;" but just how you are to look out for it is not apparent. Women tell their husbands and children to wrap up in warm clothing, as though a microbe cares how one is dressed.

An intelligent understanding of the composition and the functions of the body will enable any one to interpret the signals or calls sent out, and immediately supply the material that is lacking. One should understand that *"All* things work together for good," that the amount of food should be regulated, plenty of rest and the right amount of exercise be taken at the right time. Then a combination of the twelve cell-salts found in the body should be taken systematically in order to supply the deficiency in the human laboratory.

When the deficiency occurs, the patient, not being at ease, imagines that something must be the cause of the so-called pain, fever or unrest; and, in this, he is encouraged by the Doctor, who coins a Latin or Greek word to give standing and personality to the myth. The condition of not-at-ease arises because the blood, not being perfectly organized blood, does not properly feed and nourish the nerves, muscles and other tissues of the body, and, then, a call or dispatch is sent to the throne of understanding, asking that the lacking principle may be supplied.

These words, asking for the workmen to build new tissues into the human organism have, through the grossest ignorance, been given names in Latin or Greek, clothed with the tinsel of so-called scientific authority, and the people have been called upon to recognize and bow down in fear before the devils thus let loose.

An irreverent writer not long ago said that the romances of the present day were written by scientists, and that Rider Haggard, with "She" and "King Solomon's Mines," was but a dabster in comparison. A short time ago the *Homeo-*

## Theory of Disease

*pathic Envoy* said: "The march of science is preceded by an ever-increasing horde of Greek and Latin terms which stupefy the brain of the unscientific who seeks to comprehend them."

Among the latest of these words to carry dismay to the learned and unlearned is "phagocyte." Occasionally the editor of a daily paper is seen struggling with this foreigner, as though he knew all about him, and in late medical journals he may have been seen stalking across the pages of some of the heavyweights. Of course, no one would display his ignorance by asking what sort of a thing a "phagocyte" is; so we will look up his pedigree in the lexicons. The word "phagocyte" is derived from the Greek "Phagein," to eat, and "cytos," a hole or cavity, and really means an eater with capacity—a ravenous eater—but, according to the scientific gentleman, a "Phagocyte" is an eater of bacilli. Now, no doubt, the grand idea becomes apparent: turn the phagocytes loose in the system and let them run out the bacilli, as ferrets run rats out of a barn. One learned editor of a great daily, in the early days of lymph, suggested that it was a phagocyte, and discoursed most learnedly on the vast fields this new discovery of science opened up, and speculated as to whether each breed of bacilli had its own phagocyte or whether one breed of phagocyte could sail in and wipe up the floor with the most ferocious bacilli.

One question, however, has not yet been considered, and we respectfully suggest that science turn its light in that direction. After the pugnacious phagocyte has cleaned out the bacilli, what creature shall we turn loose in our insides to fight the phagocyte?

For the benefit of those who wish to know what the phagocyte may be called upon to contend with, I will mention some of the latest discoveries in the realm of the microbe: Man's liver may be infested with the terrible *distomum hepaticum,* while that of mutton suffers from *distomum lanceolatum.* The rabbit whisks about with *coccidium oviform* in his inwards, while man and cow both furnish habitation for the gentle *echinococcus polymorphus.* Man alone seems to have the distinction of entertaining the aristocratic *bothrioces phalu latus* and that free- booter of the highway, *ankylos-*

## The Chemistry and Wonders of the Human Body

*tomum duodenale.* The sporting couple, *dochimus tringocephalus* and *stenocephalus*, seek the society of hunting dogs only; first cousins to these, but rather more aristocratic, are the brothers, *scleorstoum hypostomum* and *tetracanthum*, who ride in horses; while the disreputable family on whom all well-regulated microbes look down, that is, the *strongylicontortus, fillicolus, strigocus* and *retortor formis,* dwell amidships in goats and such like plebeans. But the most noted of all is the musical *dochmius atenocephalus,* the intimate of the cat, which, as further research will no doubt reveal, must be the microbe of all attempts to reach high C.

But Virchow's researches completely overthrow the germ or microbe theory and clearly prove that disease is caused by a lack of some constituent of the blood at the part affected and not by germs or bacilli.

The human system can use its constituent parts only; the cells are not fed—they feed themselves. They reject what they do not need. It cannot be forced upon them, except to the detriment or death of the body. Our vital forces are at once set to work to rid the system of anything and everything that does not belong to our organism and will not assimilate with blood, bone, muscle or other tissues.

Calomel, quinine, aconite, belladonna, salicylic acid, opium and the thousand and one poisons used in the "regular" medical practice are not constituent parts of the blood—are not found in the human organism, and, when taken into the system, set up their own action for the abnormal condition called disease and are worse than the disease itself.

Calomel does not cure; it simply sets up a diarrhoea in place of constipation. Opium does not cure; it simply causes paralysis of the nerve centers, in place of neuralgia. Those who take poisons and yet recover, do so in spite of both the disease and the drugs. The normal condition is restored through the natural process. The so-called medicines have no part in the restoration. No improvement can be made on the human organism in this respect. The constituent parts of our bodies, when perfectly balanced, keep all in harmony. When an abnormal condition arises, harmony can be restored by restoring the balance, but not by introducing a poison into the system. The disease may be changed to one

## The Fallacy of the Germ Theory of Disease

that manifests itself in a different manner, but the patient is not cured. The word poison has but one definition, that is, an agent which, when taken into the stomach or blood, produces either disease or death. Therefore, by no possibility can poison cure.

When the microscope first revealed the fact that there is no inert matter—that all so-called matter is life in operation, even in the crystal and diamond—the scientists were frightened out of their wits. They saw God face to face, named Him "microbes" and tried to kill Him with carbolic acid.

But "He that sitteth in the heavens laughed and had them in derision." Carbolic acid is one expression of "Omnipresent Life"—microbes are another expression. Add the two and you still have the life.

The true thing alone is orthodox and no length of time can sanctify error. Not many years ago, the State Board of Health of Louisiana caused cannon to be fired in the streets of New Orleans, expecting the concussion to *kill the germs of yellow fever.* Later, the Seers of medicine declared that mosquitos were the sole cause of this disease.

It is claimed by the adherents of the germ theory that so-called malarial conditions are caused by germs.

Dry air cures ague. Cold weather cures ague. Sodium sulphate (*natrum sulphuricum,* or sulphate of sodium) in 3rd X-one of the cell-salts of the blood—cures ague. So, then, these must all be Royal Germ Killers. No, they supply deficiencies.

Ague is caused by an excess of water in the blood and *dry* air (cold air is dry air), furnishes an extra amount of oxygen to the blood through the lungs, and eliminates the excess of hydrogenoid gases or water.

Sodium sulphate molecules eliminate an excess of water from the system. Each molecule has the chemical force to carry two molecules of water. *No one ever has ague whose blood is properly supplied with* SODIUM SULPHATE, NO MATTER HOW MANY GERMS OF MALARIA MAY ASSAIL HIM. The *Homeopathic News* published the following editorial on the germ theory in 1892:

"It is to be hoped that no intelligent homeopath believes

in a microbic origin of disease. It may be a matter of interest to the physician to study the bacteriological accompaniments of a given malady—if the malady has any; but the idea, that the fullest knowledge of the particular bacillus that may be found in the body of the subject of a given disease is of practical benefit to anybody, is a mistake.

"For whether we empirically administer drugs, hoping to cure disease, or prescribe experimentally for the destruction of a bacillus known to us microscopically, is all one. Give us a specific for cholera, and we care not whether our specific cures by destroying certain bacilli, or by producing blood changes, or in any other way. Where there is no guiding law for the cure of disease, it is all try, try, let the cause within the organism be animal, vegetable, or mineral, known or unknown.

"But we do not believe that, if bacilli peculiar to certain maladies have been found, they are the cause of the diseased state, that they accompany, any more than we believe that the leaves on a tree are the cause of the existence of the tree.

"When a theory of the causation of disease is backed by names universally admitted to be as great as those that endorse the germ theory, intelligent men are willing to investigate it. This we have done most thoroughly; and we believe that theory to be erroneous from top to bottom, and from first to last. We have never found a particle of evidence that bacilli have been discovered—in the sense pretended by their 'discoverers.' They have undoubtedly been 'faked' to a very great extent, beyond doubt intentionally on the part of the chief promulgators of the germ theory. The medical profession cannot be blamed, if its members very much doubt whether those gentlemen themselves believe in their own theory, or in their own 'discoveries.' We have learned to know, that disease is not an entity, but a condition produced by deficiencies, and that germs are a product of these conditions and do not cause the conditions."

The remark is frequently heard that the baby *"nursed"* its sore throat or bad cold from its mother. The statement is not only dogmatic and crude, but in the light of Biochemistry, is un-scientific.

The new pathology claims to be able to scientifically dem-

## The Fallacy of the Germ Theory of Disease

onstrate the fact that so-called disease is simply a condition, and not an entity that may be transferred from one to another. Therefore, the expression, "Caught it from its mother," cannot be correct.

But then, the question arises, "How shall certain facts be explained"? No one will venture to deny, that nursing infants are very liable to suffer from the same symptoms that manifest themselves in their mothers; and, when we take issue with the race belief as to the *modus operandi,* by which the condition of the mother appears in the child, it is meet that we should offer our reasons and suggest the true cause of the phenomena.

In order to make the matter clear, I will offer an illustration: Suppose a child, say five or six years of age, should be fed on a certain kind of grain, known to be deficient in phosphate of lime, and should, as a consequence, suffer from the disease or condition known as rickets, *rachitis,* admitted by all schools to be caused by a deficiency of the lime salts of the blood. In such case, no one would maintain that the grain gave the child the rickets, or that it caught it from the grain, but rather, that the grain, being deficient in lime, but not in albumen, furnished the blood with a sufficient quantity of *organic* matter, but not enough mineral or inorganic material to build up true bone structure.

Now for the application. Before the mother (or any one else) can have a cold or sore throat, there must be a deficiency in one or more of the inorganic cell-salts, or tissue builders, of the blood. Let us suppose the salt, that has fallen below the maximum, to be potassium chloride (kali mur), and, as a consequence, a certain portion of fibrin, not having workmen—molecules of kali mur,—to use, was thrown out by the circulation and clogged the parotid gland or tonsils, or other glands, or caused irritation to the membrane in nasal passages, or larynx, bronchial tubes, or pleura, or clogged the air cells of the lungs. In such case, is it not reasonable to suppose the mother's milk would be deficient in the cell-salt kali mur, and that the child, in accordance with the law laid down above, would also suffer from the *result of the deficiency,* as did the mother?

As to germs, or bacilli, or microbes, etc., they swarm

## The Chemistry and Wonders of the Human Body

throughout all nature. *They are Omnipresent Life in operation.* They adhere to membranes in unhealthy conditions, but do not affect healthy ones.

Decaying organic matter produces microbes that exist while the process of decay goes on, feed *upon* it and disappear with it and return to the elements from which they were materialized.

# DIPHTHERIA

LET me call your attention to the pathology of diphtheria (Greek for membrane), that scourge which baffles the skill of the regular practitioner, although his office walls are decorated with diplomas written in Latin. Venous blood fibrin amounts to three in one thousand parts; when the molecules of chloride of potassium fall below the standard in the blood, fibrin thickens, causing what is known as pleurisy, pneumonia, catarrh, diphtheria, etc. When the circulation fails to throw out the thickened fibrin via the glands or mucous membrane, it may stop the action of the heart. Embolus is a Latin word, meaning "little lump" or balls; therefore, to die of embolus, or "heart failure," generally means that the heart action was stopped by little lumps of fibrin clogging the auricles and ventricles of the heart.

In diphtheria we have a striking illustration of the effect of deficiencies. Electric changes and disturbances in the atmosphere cause a deficiency in molecules of potassium chloride.

But some one may ask how atmospheric or electric states and changes affect our cell structure, so that conditions called disease appear. It is well known that there are subtle influences which, although invisible, produce well-defined results. The barometer rises or falls while the degree of heat or cold remains unchanged. Every baker knows there are influences unseen and unfelt by him that prevent his yeast from rising. They are simply antagonistic influences.

When these influences are adverse to man, he suffers just in proportion to his inability to meet the opposing forces. The forces that injure him first act through the pneumogastric nerve, disturb the gastric juice and break up the chain of molecules of iron. This diminishes the outer circulation, the feet and hands are probably chilled, the pores of the skin close and the waste matter, the dead cells that should

## The Chemistry and Wonders of the Human Body

be cast off through these avenues, are turned upon the inner organs. It is the law of conservation of energy that motion is changed to heat; and when the machinery of our being is set more actively to work to rid the system of this waste matter, the increased circulation which follows produces an excess of heat which is called fever.

In diphtheria, which, as said before, is caused by a deficiency in molecules of potassium chloride, these salts with fibrin and albuminous substances find their way to the tonsils and thymus gland and form a plastic exudation. From the supply constantly thrown out of circulation the accumulation keeps growing until the patient dies of suffocation. There is no specific diphtheria germ.

# CHOLERA

CHOLERA is a Greek term, derived from "chole," meaning bile. Cholera is a chemical condition characterized by violent emesis, diarrhoea, abdominal pains and cramps. The alvine discharge, resembling rice water, with floculent sediment, indicates great di-sturbance in the gray matter of the brain and a breaking up of the nerve fluids; also a non-functional operation of water, which shows that sodium chloride molecules have fallen below the standard of balance, and, therefore, fail to properly control and distribute water. The chief cause of the acute attack is the breaking away of water from blood and serum.

But the primary cause is an over-supply of water in the blood, caused by an atmosphere heavily laden with moisture. Cholera does not thrive in temperature below 70 degrees, although cases sometimes appear after the temperature has fallen below 70 degrees, but the blood has become overloaded with water during the time the heat was great enough to cause excessive humidity. Persons who have not been exposed to humid air, do not yield to the disease, although they may come in contact with a cholera patient. Therefore, my centention is, that an excess of moisture (pure water $H_2O$), is the cause of cholera, and that germs, microbes, bacilli, etc., are concomitants of the chemical break in the blood and nerve fluids. Pure water may thin the fluids of the liver until fatal results are produced.

The so-called cholera bacillus lives on the heteroplasm caused by the molecular break in the chain of the mineral salts in the blood. Sodium sulphate regulates the amount of water in the blood by its chemical power to eliminate the surplus water. But when the air is overcharged with aqueous vapor, sodium sulphate molecules often become overworked in their efforts to eliminate water, and a deficiency arises, leaving the "water logged" system in the condition known as cholera, yellow fever, or malaria (bad air), according to the degree of the deficiency coupled with planetary aspects at the particular time. In most instances the person with too much water in the blood will simply suffer from the condition known as chills and fever, or malaria. The chill is a spasm of the muscular nervous and vascular system, making a supreme effort to wring out and throw off the surplus water in the blood that would have been done by the catalytic action of sodium sulphate, had there been a reinforcement of that cell-salt to equal the amount of water taken into the arteries through the lungs by breathing air over-charged with humid vapor.

The heat (called fever), that follows a chill, is but the result of rapid circulation (friction), caused by nature's efforts to carry oxygen to all tissues of the body to supply the deficiency caused by the spasm or chill.

If the amount of water in the blood should be excessive, and planetary angles favorable, the fluids of the liver and pancreas break away. The liver first empties its contents, and the discharge is colored with bile, but, later on, the fluids are clear, or like rice water, which indicates water and nerve fluids. All other outlets of the body seem closed except the intestinal tract. The pores close and the urinary secretions become dried at their source.

In 1852 Peyton wrote, in a treatise on cholera: "Very remarkable results have been found to follow the injection into the veins of a dilute solution of saline matter resembling, as nearly as possible, the inorganic salts which have been drained away."

During a cholera epidemic in 1877 the famous Dr. Koch went to India to search for the cause of that dread disease. A microscopic examination of the cistern water (it had never been examined before) revealed bacilli, (Latin for little sticks), and Dr. K. sagely concluded that he had found the cause of cholera. The discovery was telegraphed to "earth's remotest bounds," and the fatted calf was killed and eaten, while the medical world held "high jinks;" but they could not find anything that would kill the germs without being quite as fatal to the patient as cholera. After colder weather came and the cholera epidemic subsided, some doc-

*Cholera*

tors who were skeptical about the microbe nonsense (quacks, I suppose), went to India to search for the dead germs, thinking, of course, that as there were no more cholera, there could be no more germs; but, on examining the cistern water, they found the bacilli lively and in good health, but they refused to bite any more. It seems that these microbes only got real hungry during the hot weather.

# PNEUMONIA

THE following, on pneumonia, was taken from the Biochemic System of Medicine, published in 1894: "No abnormal condition with which suffering mortals are afflicted has such terrors for the average physician as pneumonia, unless it be typhoid or typhus fever. The New York Medical Record, of late date, contained the following: 'Pneumonia is attended at the present day with an ever-increasing mortality—so high, in fact, as to constitute a reproach to medical science.'

"A prominent physician of Philadelphia wrote as follows: 'The mortality of pneumonia in Philadelphia has increased, and is greater today than it was thirty years ago.'

" 'The errors of thirty years ago have been intensified. Still larger doses of nauseating drugs have been administered, and the local treatment has increased in severity.' The microbe theory, the fallacy of the age, has piled error on error, until the wonder is that any one recovers from pneumonia under the treatment. The trouble all along has been the failure of the medical profession to understand just what causes the condition of: 'not at ease,' in pneumonia.''

The medical text-books and dictionaries will tell you "that pneumonia, or lung fever, is inflammation of one or more lobes of the lungs;" but make no attempt to explain what inflammation really is or what produces it. These books tell us that certain anatomical changes take place; that firm or solid exudation is found in pulmonary alveoli, but do not tell of what the exudation is composed or how it reached the air cells. We are graciously informed that "there is intense congestive hyperaemia;" that "there is red hepatization, in which the lung is bulky, heavy and airless, its red tint due to extravastated corpuscles and distended capillaries or gray

## Pneumonia

hepatization, due to decolorization of the exudation and pulmonary anaemia or colliquation and resolution."

The above may be very learned, but the student cannot help but wonder what the cause of it all is, and what on earth he will do to prevent it.

Biochemistry alone explains the cause of the abnormal condition called pneumonia and offers the cure.

Atmospheric electrical changes so operate on the human system as to cause deficiency in iron and other inorganic mineral salts of the blood, which lowers the vitality and causes the pores to close. The waste matter, the exudation from the skin, is then turned inward and seeks an outlet. The fluids of the body, water, albumen, etc., serve as carriers for the effete matter.

Of course the circulation is increased, for two reasons: (a) to carry off the decaying organic matter; (b) because of a deficiency in iron the blood is poorly supplied with oxygen, which, as is well known, has an affinity for iron; and the rapid motion is nature's effort to make the limited iron supply do the work of the maximum amount.

If, in getting rid of the waste, nature directs it to the lungs, and this causes injury to lung structure, as the decaying, vitiated, organic matter surely will, if in sufficient quantity, the medical profession has been pleased to name it pneumonia. Catarrh, bronchitis, etc., have the same pathology.

The decaying organic matter, the heteroplasm, deposited in connective tissue and membranes during the inflammatory stage, of course, must be gotten rid of, but the circulation will attend to it, if the tools are furnished to work with. While this work is being carried on, there will be disturbances, coughs, etc., but the only rational way to do is to supply the blood with the vital principles needed, so that the work may be done.

# DIABETES MELLITUS

SEARCHING for a rational explanation of the cause and cure of diseases, the student is an object of pity. Honestly desiring to obtain data whereby he may benefit his fellows by restoring the sick to their normal condition, he is met at every turn by vague conjectures, wordy explanations that do not explain, and reports and clinical cases that simply show the dense ignorance of the age in regard to the true pathology.

Suppose the medical student wishes to learn the underlying cause of the disease or condition called diabetes mellitus (I pass sugar), reads the latest thoughts of the day, what does he find? Ebstein will tell him that "the diminished elimination of $CO_2$, which is characteristic of diabetes, is the cause of the large sugar proportion, because, in health, the action of the diastatic ferment upon glycogen is held in check by the $CO_2$." Catani declares that, "the diminished $CO_2$ is the result and not the cause of the diabetic; there is less $CO_2$ because there is less combustion of gycogen." And again the student must be content that out of 1004 cases of diabetes mellitus 837 were males and 167 females. Arthur Flagg informs us that in order to have true diabetes we must not only have a rapid outflow of sugar from the liver, but also an increased formation of glycogen in the liver; but as the student is not hunting for diabetes, but for the cause and cure, he passes on in disgust. Professor James Anderson: "We must go back to the nervous centers, central and local, by which this action is controlled. It must be remembered that there is both defect and excess, both paralysis and stimulation, underaction and overaction, at the same time."

Imagine a young doctor writing a prescription after reading the above.

Dr. G. Arthand thinks "that by irritating the centrifugal

## Diabetes Mellitus

vagus nerve, different varieties of diabetes, such as insipidus, agoturic, and glycosuric, can be produced." But I notice he fails to tell us how to cure these dreadful specimens of diabetes. The student will likely be careful not to irritate the centrifugal vagus nerve.

The French scientist, Lancereaux, thinks the pancreas must use its influence in some way to prevent diabetes, as he extirpated the pancreas of a dog, and "diabetes immediately set in." It will probably seem strange to the student that nothing worse than diabetes "set in;" but as the dog only lived a few days, I suppose nothing else had time to set in.

Albert Robin wisely concludes: "That diabetes is essentially characterized by the production of an abnormal amount of sugar through the exaggerated function of the hepatic cells." Well, what of it? Nothing. But Robin is not altogether correct in his learned conclusion, by any means; for we are informed in the "Annual of the Medical Sciences" that Jules Worms opposes the views of Robin and claims that "the theory of hyperglycogenesis has not been proved; because the experiments of Robin were made upon patients highly agoturic and highly glycosuric." Let us hope the patients were not aware of their condition. But, at any rate, that settles Professor Robin, and it ought to; he had no business to experiment on patients in such a condition.

Meyer, of Naples, Italy, says that, "in eight cases of diabetes studied by him he has always been able to ascertain anatomical alterations more or less deep in one or more organs of the digestive system." Well, what of it?

P. Ferraro finds that "in the lungs of diabeties are morbid changes not due to bacilli." This is strange in the age of the festive bacilli and all-pervading microbe. I do not see how they are ever to be excused.

But, enough. What has Biochemistry to say about diabetics? Lactic acid is composed of carbonic acid and water, and must be split up on its way to the lungs. This is done by the catalytic action of sodium phosphate in the blood. Any deficiency of sodium phosphate will cause a disturbance in the water in the system by allowing an excess of lactic acid to accumulate. Nature, in her effort to eliminate the water, produces the symptoms called diabetes.

## The Chemistry and Wonders of the Human Body

But while a lack of sodium phosphate is the principal cause of diabetes, the chief remedy is sodium sulphate, because it regulates the supply of water in the blood. Sodium sulphate also gives off oxygen, so necessary for the process of the decomposition of sugar, and thereby prevents its reaching the kidneys as sugar, and also thins to its normal consistency bile that has become inspissated from a lack of sodium phosphate.

If a case of diabetes has advanced to any considerable degree, the kidneys will have become inflamed by the lactic acid and sugar that passes through them. This injury to the tissues of the kidneys calls upon the red blood corpuscles for iron phosphate, which will, in most cases cause a deficiency in that inorganic salt. Nature, in her efforts to supply iron, will probably draw on the nerve-fluid, potassium phosphate will be too rapidly consumed, and the patient suffers from nervous prostration.

The treatment, therefore, for diabetes mellitus is the phosphates of sodium, iron and potassium, and the sulphate of sodium. For the great functional disturbance of nerve centers caused by the demand made on the blood for the potassium phosphate, producing sleeplessness and voracious hunger, Kali phos is the infallible remedy. It establishes normal functional action of the medulla oblongata and pneumogastric nerve, which latter acts upon the stomach. For the great thirst, emaciation and despondency, give Nat. mur. It equally distributes the water in the system and quickly restores the normal condition. Diabetic patients should eat very little, and use no sugar. A diet of fat meats or greasy food cannot be beneficial, for the very important fact that it overworks the liver, causes a deficiency and consequent thickening of bile and mucus, and sometimes a crystallization of cholesterin in gall-duct, which gives rise to symptoms called hepatic colic, jaundice, or bilious headache.

# SYPHILIS

FOR many long, weary years, people thought the sick had a mysterious poison in their blood that no one could see, smell or taste; but, as they were sick, it must be there. Were they not afflicted? Look at the eruptions on the skin, or an exudation from a sore or some orifice of the body! Is not that proof positive that the poison is doing its work? And so they set about to counteract the terrible materies morbi by dosing the patient with poisons.

The ignorance of the world in regard to the pathology of syphilis, is appalling. The general belief and treatment of these symptoms, caused by a deficiency in the blood, has caused more misery than war. It has driven men and women insane. It has darkened the lives of children, and hung a pall of blackness across the future. It has been the skeleton at every marriage feast; and, chattering its senseless jargon, it has impeded the onward march of man. Like the sword of Damocles, it hangs above the social world and proclaims itself king amid the wails of heart-broken thousands. As false as our past ideas of Satan, Science has bent the knee, and acknowledged it master. And what is this THING that I call false and unreal? A deficiency in potassium chloride, first causing an outflow of fibrin, which, reaching a very delicate membrane, plexus of nerves, and minute system of infinitesimal blood-vessels, sets up irritation by abrasion or pure physical injury.

When the Contagious Disease Act was before the British Parliament, the chairman of the Government Commission appointed to investigate as to the extent and danger of syphilis, reported: "The statements made by the promoters of the bill are largely overcharged, and too highly colored. The disease is by no means so common or universal, I may say, as is represented in that article; and I have had an opportunity, since I was summoned here, to appear before the College of Surgeons."

John Simon, F.R.S., for thirty years hospital surgeon, and now medical officer to the Privy Council, writes, in his official capacity: "I suspect that very exaggerated opinions are current as to the diffusion and malignity of syphilis."

The late Professor Syme (English surgeon) said: "It is now fully ascertained that syphilis does not give rise to any of the dreadful consequences which have been mentioned, *when not treated with mercury.*"

The British and Foreign Medico-Chirurgical Review says: "The majority of those who have undergone the disease live as long as they could otherwise have expected to live, and die of diseases with which syphilis has no more to do than the man in the moon. Surely 455 persons suffering from syphilis, in one form or another, in a population of one million and a half (one in 3000), cannot be held to be a proportion calling for action on the part of the government."

Holmes Coate, F.R.C.S., surgeon and lecturer at St. Bartholomew's Hospital, says: "It is a lamentable truth that the troubles which respectable, hard-working married women of the working class undergo, are more trying to the health and detrimental to the looks than any of the irregularities of the harlot's career."

W. Burns Thompson, F.R.C.S., says: "The statements in that act seem to me to be gross exaggerations."

Surgeon-Major Wyatt, Dr. Druitt, health officer, and Dr. Acton, specialist, J. Hutchinson of the British Medical Journal, and Ambrose Pare of the University of Paris, are all on record that the ravages of the disease have been exaggerated a thousand-fold. Notwithstanding the fact that the highest medical authorities hold that syphilis is not a serious disease when let alone, and not treated with mercury; notwithstanding the fact that mercury produces violent gastralgia and diarrhoea, rots the bones and teeth, and leaves the victim in a worse plight than the disease for which it is given, could we see doctors, with legal diplomas, dosing their unfortunate victims with that baleful drug, as though it contained some mysterious potency of life.

Bockhart's investigations with injections of mercury,

## Syphilis

show that for forty-eight hours the pains were excessive, and many patients die under the treatment. Two cases were reported by Balzer, where death occurred from the treatment.

Lesser, after experimenting on five hundred cases with mercury, declares that once in the system no one can forsee what evil it may produce, while powerless to help the patient get rid of the mercury.

The Biochemic materia medica treats syphilis as it does catarrh.

The salts should be used both internally and externally. For ordinary cases a solution of Kali mur. makes an excellent wash. The parts should be bathed frequently and kept perfectly clean.

Bear in mind that the exudation is fibrin and other organic matter thrown out of the circulation during the first stage, because of a lack of potassium chloride salt in the blood. Supply this deficiency by repeated doses of this inorganic salt, and the supply of the irritating substance is cut off. Syphilis is not a thing in the blood, but a lack of a thing.

# SMALL-POX

SMALLPOX, or variola, caused, according to the teachings of Biochemistry, by a large amount of organic matter becoming non-functional, and in being thrown out through the skin causes the papular eruptions. This decaying organic matter, this heteroplastic accumulation, within two weeks passes through periods of vesiculation, pustulation and incrustation. Whatever may be said about the origin of smallpox, one thing has been proved to the satisfaction of every Biochemist, i. e., that there must first arise a deficiency in potassium chloride, the cell-salt which controls fibrine, before the material which causes the lesion, the eruptions, is thrown out of the vital circulation, and reaches the skin. The real pathology of smallpox and measles is the same. Smallpox may be "catching," but there must first be a deficiency of certain salts in the blood of the one who "catches" it.

# GONORRHOEA

The fact that the successful Biochemic treatment of gonorrhoea is the same as the treatment for catarrh, proves the cause to be the same. Of course it is hard for those who have been educated in the old pathology to believe that a simple deficiency in potassium or sodium chloride, or calcium phosphate, can produce the symptoms classified under the heading, gonorrhoea; but when the symptoms readily yield to a treatment that simply supplies these salt and thereby prevents further escape of organic matter, one is led to exclaim, that facts are better than theories. For an explanation of the *contagious* nature of the gonorrhoea exudation, the reader is referred to the aticle on syphilis, and earnestly requested to read and reread it carefully; for in it he will find the true pathology of gonorrhoea.

# ERYSIPELAS

ERYSIPELAS is derived from two Greek words: *crythros*, red, and *pella*, skin. This disease has been so named because it generally extends gradually to the neighboring parts, and affecting not only the superficial, but, in many cases, the deeper structures, as well.

The Biochemic pathology of erysipelas is as follows: The inflammation is caused by an accumulation of both organic and inorganic matter. The organic matter in the human organism is useless in the absence of a proper supply of the mineral salts of iron, sodium, lime, magnesium, silica and potassium.

Whenever a deficiency occurs in any of them, certain organic matter, albumen, fibrin, oil or sugar, becomes non-functional or inert; because the workers are not present to take them up and organize them into the structure of tissue. Erysipelas, then, is not a *thing*, or a specific poison or microbe, but simply an injury to the skin caused by an accumulation of organic matter at a certain point on account of a lack of one or more of the mineral salts of the blood.

A very little investigation will prove that organic matter thrown out of the blood, causes the abrasions, pimples, "red skin," herpes, eczema, etc. When the fibrin, oil, albumen, etc., becomes non-functional, and is thrown out of the vital circulation, it is an irritant. If this irritating matter, because of the quantity, and also the decomposition which at once sets in, should reach the nasal passages, it is called catarrh. If it reaches the lungs, a cough is produced to get rid of it; but the first cause of these conditions is always a lack of the inorganic mineral workers that diffuse the organic material through human blood, that keeps it in proper consistency, and uses it to build up the human structure and constantly keep the waste supplied.

## The Chemistry and Wonders of the Human Body

The question has been asked: "Why does this organic matter produce different results?"

Why should it at one time break through the skin and cause itching hives or vesicles, at one time filled with yellow lymph, drying quickly only to form again, at another time, vesicles with clear, watery contents, or at another produce a "red skin" with inflammation and nervous prostration?

I answer that each of the twelve inorganic or mineral salts has its special sphere of action. One works with albumen, another with fibrin, another with water, and so on; and the number and particular combination of these salts that chance to fall below the standard, determines the kind of disturbance in the cells. A discharge of albumen, oil and sugar will cause a different symptom or a discharge of a different color than a discharge of water, fibrin and sugar. So, when the different combinations of organic substances reach the skin, they cause different phenomena. The difference in these combinations is caused entirely by a deficiency in the inorganic cell-salts that control the particular organic substances which are doing the mischief.

There can be but one true theory of disease and its cause. Biochemistry offers a rational explanation. Has there ever been an explanation offered before that even had the semblance of truth and reason?

## INTERMITTENT FEVER

WHAT a handy word is "Malaria" for the average doctor. When a person feels badly or complains to the doctor of biliousness, and asks the cause, the answer comes, without a moment's hesitation: "Oh, you have malaria." And the patient goes away satisfied that *it is* the malaria, although neither he or the doctor he consulted have the slightest idea what malaria is. Malaria simply means "bad air," and the medical world has taken it for granted that bad air in some way causes chills. The idea, or belief, arose from the fact that chills or fever and ague, prevail to a greater extent in the latter part of summer or autumn, when stagnant pools abound, or the vegetation about ponds, swamps, lakes or water-courses begin to decay.

Experience has proved that those living on the banks of the Hudson, where it runs between the banks from one to three hundred feet in height, are subject to malarial conditions as well as the dwellers in the Mississippi Bottoms.

The Biochemic pathology of ague is, it seems to me, clear, explicit and unanswerable. It is as follows: Hot weather causes a rapid evaporation of water, and it is held in solution in the atmosphere, and is, of course, taken into the blood through the lungs by the act of breathing; thus the blood becomes overcharged with water. Sodium sulphate, one of the mineral salts of the blood, has an affinity for oxygen, and oxygen has an affinity for water, and thus, in health, the surplus water may be eliminated from the blood, and the proper balance maintained; but if, from any cause, the molecules of sodium sulphate fall below the standard when there is much moisture in the air, as there always is in hot weather, the blood does not receive enough oxygen through the process of respiration to eliminate, or carry off, the extra amount of water. Hence, anyone who lacks the even balance of sodium sulphate is liable to "malarial" conditions. The water iself in a swamp is as pure as any water when it is separated from the organic matter, i. e., the vegetable growth found in the swamp. This vegetable growth, when decaying, may emit a disagreeable odor, but has no possible connection with the "chills." A chill is a spasm or violent exertion made by the nervous, muscular and vascular system to throw off the surplus water. It is always followed by increased circulation and perspiration, in which the excess water is disposed of. The reason the chills or spasms are about forty-eight hours apart, is because it requires about that time to again overcharge the blood, through the process of breathing, when sodium sulphate is deficient.

Dry air will always cure ague. The air is always dry in cold weather. Cold weather cures ague. An ague patient only has to ascend a high mountain to the dry-air strata to be immediately cured of chills.

The following will further explain the pathology: "If sodium sulphate is deficient in the blood, sugar is thrown off through the kidneys, the result being diabetes mellitus" (Latin for "I pass sugar"). In intermittent fever, known also as ague, chills and fever, malaria fever, the heat is caused by the increased circulation incident to the effort to rid the blood of excess of water. The quantity of water in the blood corpuscles and the blood serum is increased, and consequently the quantity of oxygen taken up by the blood is diminished. Sodium sulphate promotes the removal of the excess of water from the organism. When, by its action the proportion of water in the corpuscles has been reduced to the normal condition, the corpuscles are again able to take up the full amount of oxygen and distribute it to the tissues. As the tissues are in this way brought back from their pathological to their normal physiological condition, they are enabled to remove from the organism the cause of the ague, be it marsh, gas, bacteria, bacilli, microbes, or germs. Dry mountain air, which is rich in oxygen, cures ague spontaneously, because the organism takes up a large amount of oxygen and disposes of much water by evaporation. Evidently, dry air is death on germs and should be named the royal germ-killer. And here it is proper to remark that the

## Intermittent Fever

excess of water in the blood is obtained entirely from the air, and that any amount of water we may drink can have no effect upon it. Therefore, the chief remedy for all malarial conditions is Natr. sulph. Those who have been afflicted for some time will probably need Kali phos., also, and the fever always calls for Ferr. phos.

In cases of drowsiness, especially in the forenoon, Nat. mur should be given, two or three doses daily. The reason quinine will frequently break up chills is: First, nature has only one vehicle to carry off waste matter or any substance not needed. The moment any foreign substance is taken into the organism the fluids of the body (water) are called upon to wash it away; thus the amount of water is reduced and the chills temporarily checked. Second, quinine contains a small per cent of both ferrum phosphate and sodium sulphate. But sodium sulphate and ferrum phosphate are the remedies and not the quinine. It is much better to furnish the blood with the natural tools to carry off the surplus water, which are sodium sulphate and oxygen, than to incidentally reduce the amount of water in the blood by calling upon it to eliminate a poison from the system.

Deafness and blindness are frequently caused by the prolonged use of quinine, because it eliminates too much water and dryness of the membrane is the result. Should this occur in the structure of the middle ear or lining membrane of the eustachian tube, partial deafness will be the result. If a dryness, or deficiency in water should be caused in any of the delicate membranes of the structure of the eye, the sight will be impaired. Which is worse, to have chills, or to be blind and deaf? Cell-salts build up; they do not tear down.

# TYPHOID FEVER
## From Typus, Meaning Stupor

ONE who fully understands the Biochemic pathology, will see at a glance that like symptoms call for the same treatment, under whatever name they may appear. There are certain atmospheric and electrical conditions that tend to weaken the elastic fibres of the skin and connective tissue, and thereby close the pores.

The typhus deposit, a peculiar substance of new formation, found in the areolar membrane, between the mucous and muscular coats of the patches of Peyer in typhoid fever, is not a specific poisonous substance.

Analysis has proved it to be the waste matter, the dead cells of the body, together with albuminous substances in a state of ferment and decomposition, which is the seat of the local trouble.

Nature, in her effort to eliminate the accumulation of decaying organic matter, increases the circulation and motion produces heat. Science may call the heat fever, but heat will do just as well.

There is another cause of the rapid circulation. A deficiency in the cell-salt phosphate of iron must have occurred before the epithelial cells became so weakened that the pores of the skin refused to perform their natural functions.

Iron molecules, it is well known, are carriers of oxygen, and when they fall below the standard, the blood is not able to carry enough oxygen to all parts of the body to properly sustain life without increasing its speed.

Seven men cannot do the work of ten, except they work faster than ten.

The foul odor that is characteristic of this disease is caused by phosphureted hydrogen set free from the nerve fluids because of a lack of the cell-salt potassium phosphate.

## Typhoid Fever

We now see why this salt is the great remedy in all malignant conditions or where decomposition, or bad odors, appear.

In the first stage of typhoid, Ferr. pho., Kali phos. and Kali mur should be given. If the skin be dry, with chilly sensation, Kali sulph. should be given instead of Kali mur. This treatment, with copious injections of hot water for the bowels, will nearly always break up the condition in two weeks, if taken at the very onset.

Where the disease is running its course, give Kali phos. as the chief remedy. Ferr. phos. for fever, except afternoon fever, which calls for Kali sulph. Should a green, alvine discharge appear, give a few doses of Nat. sulph. When patient wanders in mind, with low mutterings, or bubbles of saliva on tongue, or great languor and weariness, give Nat. mur. in alternation with Kali phos.

# YELLOW FEVER

WHEN the real cause of an evil is not definitely known, it is attributed to some malignant source. When a person has an attack of diarrhoea, it is supposed to be caused by over-eating, or indigestion, or a sudden change of weather, or severe hot weather; but if the condition be so severe as to cause death, the world at once jumps to the conclusion that it is cholera, and caused by the cholera germ. If one has a simple fever of a mild type, the doctor calls it bilious fever; but should the conditions antagonistic to health and life be great, if the skin turns yellow, and vomiting of a dark-brown substance set in, which quickly kills the patient, the scientific world declares a deadly germ to be the cause of the disease.

To show the ignorance that has prevailed in regard to the cause of yellow fever, I only need cite the fact that a few years ago the State Board of Health of Louisiana caused cannon to be fired in the streets of New Orleans, expecting the concussion to kill the germs of yellow fever and thus "stamp out the disease." Yellow fever and cholera are almost exactly the same in their pathology and treatment, and they both require a temperature of about 72 degrees, and are not found beyond 48 degrees of north latitude.

In yellow fever a vitiated condition of the bile causes an inflammation of the colon, which closes or partly closes the outlet. The excess of water in the blood, caused by a high temperature in a low district near the ocean, or large bodies of water, is the primary cause of the disturbance in the liver. The bile and the fluids of the liver and pancreas are then distributed through the body by absorption; the pores become closed and the kidneys cease, in a great measure, to perform their functions; and the secretions of the body, thus held in check, rapidly decompose and produce the yellow skin and black vomit. The fever is simply heat, caused by

## Yellow Fever

nature's effort to supply the various parts of the body with enough oxygen and inorganic material from the limited amount on hand. All fevers are produced by this same cause; but other causes, different states of atmosphere and planetary and electrical conditions, produce different symptoms, hence arises the multitude of names.

Nat. sulph. will prevent yellow fever by furnishing the blood with the tools to eliminate the excess of water taken in through th lungs by breathing an atmosphere heavily charged with moisture.

# LESSON I.

THE chemical formula and Physiological action of the Cell-Salts of the human Organism.

As bone is the foundation of the animal structure, I will commence with the bone builder.

Phosphate of Lime—Synonyms: Calcarea Phosphoricum, Calcium Phosphate. Formula $Ca_3 Po_4)_2$.

Phosphoric acid, dropped in lime water, precipitates this salt in crude form. Let the student bear in mind that this lime salt, as well as all the others, must be triturated with sugar of milk up to at least the third decimal, or potency, in order that the molecules may become separated from the mass to the extent that they can be take up by the delicate mucous membrane absorbents of the stomach and intestinal tract and thus enter the blood vessels.

As to the best decimal potency to give, to supply deficiencies in the mineral salts, there is quite a difference in opinion. Many of the homeopathic physicians contend for higher potencies, from sixth up to two hundredths, while biochemists generally use the third and sixth.

Bone tissue consists of about 57 per cent lime phosphate. The lime salt has chemical affinity for albumen. While there is a certain degree of affinity between each of the cell-salts and albumen—albumen being the base of all organic matter—the operation of calcarea phosphate with albumen is greatest. The salt chemically unites with albumen, carries it and uses it as cement to build bone tissues. Bone also contains carbonate of soda, magnesium phosphate and sodium chloride, but lime phosphate is the chief builder of bone tissue, and it follows, as a logical sequence, that it is the principal salt deficient in all so-called diseases of bone structure.

The gelatine found in bone tissue is formed by the union of albumen, oil, carbon, lime and phosphate in certain pro-

## Lesson I

portion. Of course, there is a small amount of other principles in bone, such as magnesia, sodium chloride, silica, etc. When the molecules of lime phosphate fall below the normal in amount and thus fail to keep up the supply of bone material, some symptoms of bone disease manifest. Again, a lack in the proper amount of this builder, in some instances, causes an anemic condition, for bone material (lime and albumen) is the foundation of bodily structure.

Under certain conditions, dependent on deficiencies in other cell-salts, a break in the molecular chain of lime phosphate will cause an outflow of albumen through the kidneys. Why should the escape of albumen through the kidneys be named Bright's Disease? It seems that the fact of the loss of albumen in this manner was first discovered in the case of a hospital patient named Bright, and although many Browns, Joneses and Smiths die in a regular and orthodox manner from the same cause that cut off the immortal Bright, the medical profession still dignifies the disease by the original label.

The very same albumen that causes "Bright's Disease," if thrown out through the nasal passage, is called catarrh (from the Greek: to drop down).

If the albumen reaches the skin, by disintegration, or fermentation, it causes pimples, eruptions, eczema, etc. If any one derives pleasure from these names, well and good, let them use them, but chemistry knows nothing whatever about them.

A great deficiency in lime phosphate may cause albumen to accumulate in some gland and there disintegrate and flow out in pus, or heteroplasm, which is called scrofula by the old school physicians. Scrofula is derived from scrofula, Latin for sow. Maybe the ancients believed the pork eaters were more liable to the disease than were the Jews.

The lime molecules are found in the fluids of digestion and assimilation, and when there is a lack of the proper amount of these workers, the digestive juices become negative, lose their proper rate of motion, or catalytic action, ferment and thus produce gas, acid condition, etc. When lime phosphate and sodium phosphate, the alkaline salts, are deficient, acids, together with albuminous substance, may settle in the joints and thus render synovial fluids non-functional; thereby, causing pain, stiffness and swelling of the joints. Just why this chemical fact must needs have the word rheumatism tacked to it does not appear. The word rheumatism is derived from rheum—to flow out.

Biochemistry does not deal with names and effects; it deals with causes, the chemistry of life and the law of supplying deficiencies.

Medical writers of late have adopted the term albuminuria in place of Bright's Disease, which very well describes the symptom or effect, but does not hint at the cause of the symptom.

Biochemistry seeks to learn the particular food called for by pains, exudations, swelling, inflammations, etc. Nature never calls for anything that is not a constituent part of the organism demanding supply.

Blood is the base of the physical life of man, and as a man's blood is, so is his health.

A word about potencies: We are told by the analytical chemists, that a quart of milk contains only the six-millionth part of a grain of iron. An infant fed on milk receives one milligram of iron in a half pint of milk, which is only the fourth part of the above minute fraction of one part of a grain of iron. It would seem from the above that four milligrams of iron daily is sufficient to feed all the cells that are known to contain and consequently require iron. This being the case, it surely can require but an infinitesimal amount of iron to supply the links in a broken molecular chain.

With these demonstrated facts before us, high potencies are no longer vague theories and the butt of jokes. On the contrary, they are man's best endeavor to imitate Nature's processes.

# LESSON II.

SULPHATE OF LIME—Synonyms: Calcium Sulphate, Calcarea Sulphate, Calci Sulphos, Gypsum, Plaster of Paris.

Formula—$CaSo_4$.

This salt can be obtained by precipitating a solution of calcium chloride of lime with diluted sulphuric acid.

Sulphate of lime should never be used below the 6th decimal trituration.

Tissue is composed of living cells. By giving a tissue builder the deficient mineral salt in such a dose, fineness and amount as can be assimilated by the growing cells, the most wonderful and speedy restoration to healthy functions is brought about in every case of curable disease. We know that these minerals are infinitesimally subdivided in the different kinds of food we take, thus rendering them capable of being assimilated by the cells. The cells of each tissue-group receive their own special and peculiar cell-salt.

The sulphate of lime is the chief builder of epithelial tissue, or to be more exact, the chief sustainer for the chloride of potassium, as will be shown in Lesson V, is the worker in fibrine and has much to do in the formation of epithelium.

Lime sulphate furnishes the cohesive, or plaster substance to sustain the integrity of tissue. The chief symptom of disease, indicating a deficiency in the lime salt, is suppuration or the discharge of pus, which is an exudation formed by the breaking down, disintergration and fermentation of epithelium.

Lime sulphate furnishes the cohesive, or plaster, substance to sustain the integrity of tissue. The chief symptom of disease, indicating a deficiency in the lime salt, is suppuration or the discharge of pus, which is an exudation formed by the breaking down, disintegration and fermentation of epithelial cells. Lime phosphate, by its union with albuminoids,

## The Chemistry and Wonders of the Human Body

assists chloride of potassium to form epithelial tissue, or, at least, to hold it intact by its cohesive quality.

The third stage of catarrh, bronchitis, lung disease, boils, carbuncles, ulcers, abscesses or exudations from any part of the body, indicates a lack of this tissue-builder. Lime sulphate, not only sustains epithelial tissue, but, when administered in case of suppuration, it cleans out the heteroplasm from the interstices of tissue by causing the infiltrated parts to discharge their contents readily, which prevents slow decay and injury to surrounding healthy cells.

The action of lime sulphate is opposite to the work of silica (see Lesson XII), which hastens the process of suppuration in a natural manner, while the lime closes up a process that has continued too long. Thus are we made to realize the marvelous intelligence manifested in life's procession in the organism of man.

Chemical affinity is but a synonym of Infinite Intelligence in operation in the functions of man.

# LESSON III.

FLUORIDE OF LIME—Synonyms: Calcaria Flurica, Calcium Fluoride.

Chemical Formula—$CaF_2$.

This salt is formed by the union of lime and fluorine.

The inorganic salts are the workers, controlled and directed by Infinite Intelligence, which perform the ceaseless miracle of creation or formation.

It is quite as important for a student of Biochemistry to understand the process by which certain cell-salts operate to supply a deficiency as it is to know for what a particular symptom calls.

Elastic fibre, the chief organic substance in rubber, is formed by a chemical union of the fluoride of lime with albumen, oil, etc. Therefore, we find this salt dominant in the elastic fibre of the body, in the enamel of teeth and connective tissue.

A lack of this salt in proper amount causes a relaxed condition of muscular tissue, falling of the womb and varicose veins. Sometimes there is a non-functional combination of this salt with oil and albumen which forms a solid deposit, causing swelling of stony hardness; it is a sort of incomplete fibre with other lime salts and vitiated fluids of the body.

There is one particular symptom that is worthy of note in connection with the pathology of this salt. When a deficiency exists in these makers of elastic fibre in the connective tissue between cerebellum and cerebrum, the lower and upper brain, it causes groundless fears of financial ruin. It seems that the relaxed condition of connective tissue, causing a sagging of the structure of cerebellum, thereby breaks the flow of the electric or magnetic currents from the cerebrum.

The student will now see that it is exceedingly easy to diagnose disease from the viewpoint of Biochemic Pathol-

## The Chemistry and Wonders of the Human Body

ogy. No guesswork here. Go to twenty or one hundred Biochemic physicians and give the same symptoms to each, and you will get the same prescriptions in every case.

It does not matter under what name of disease a disturbance in elastic fibre appears, a study of the chemistry of life has made clear the fact that a break in the molecular chain of lime fluoride salt is always the cause of the phenomenon.

The proportion of fluorine in the human organism is less than that of iron. From analytical facts it is found that fluorine in milk is only present in decimilligrams, and yet we are confronted by the fact that this infinitesimal amount is sufficient to sustain all the elastic fibre of muscular tissue, enamel of teeth and connective tissue.

Professor Leibig, in his chemical letters, accentuates the importance of high potencies or dilutions, as follows:

"At the temperature of the hydrochloric acid, diluted with one-thousandth part of water, readily dissolves the fibrine of meat and the gluten of cereals, and this solvent power is decreased, not increased, when the acid solution is made stronger."

Why should we search Latin and Greek lexicons to find a name for the result of a deficiency in some of the mineral constituents of blood. If we find a briar in our flesh, we say so in the plainest speech; we do not say, "I have got the briatitis or sprintragia."

When we know that deficiency in the cell-salts of the blood causes the symptoms which medical ignorance dignified and personified with names of which nobody knows the meaning, we will know how to scientifically heal by the unalterable law of the chemistry of life. When we learn the cause of disease, then and not before, will we prevent disease.

Professor Valentin, the well known physiologist, says: "Nature works everywhere with an infinite number of small magnitudes, which can be perceived by our relatively obtuse organs of sense only when in masses. The smallest picture which our eyes perceive proceeds from millions of waves of light. A granule of salt that we are hardly able to taste contains myriads of groups of atoms which no sentient eye will ever view."

# LESSON IV.

**PHOSPHATE OF IRON** — Synonyms: Ferrum phosphate, Ferri Phosphas.

Formula—$Fe_3 (PO_4)^2$

Phosphate of iron may be prepared by mixing sodium phosphate with sulphate of iron. The salt precipitated by this union is filtered, washed, dried and rubbed to a powder.

The iron phosphate should not be used below the third (decimal trituration), as large doses of iron, as in tinctures, have a bad effect on the mucous lining of the stomach, injure the teeth and utterly fail to supply iron to the blood where it is needed to carry oxygen, the life giver.

One red blood corpuscle does not exceed the one hundred and twenty millionth of a cubic inch. There are more than three million such cells in one drop of blood, and these cells carry the iron in the blood. How necessary, then, to administer the salts of iron to hungry cells in the most minute molecular form.

Each one of the twelve inorganic salts has its own sphere of function and curative action. Thus we find the phosphate of iron molecularly deficient in all fevers and inflammatory symptoms.

Health depends on a proper amount of iron phosphate in the blood, for the molecules of this salt have chemical affinity for oxygen and carry it to all parts of the organism. When these oxygen carriers are deficient, the circulation is increased in order to conduct a sufficient amount of oxygen to the extremities with the diminished quantity of iron, exactly as seven men must move faster to do the work of ten. This increased rate of motion of the blood is changed to heat, caused by friction, otherwise known as the "conservation of energy."

This heat, or increase in the temperature of blood, has

## The Chemistry and Wonders of the Human Body

been named fever, from the Latin word Fevre, meaning "To boil out."

The writer fails to see any relevancy between the word fever and a deficiency in iron phosphate molecules in the blood. From Hippocrates to Koch you will not find a true definition of fever outside of the Biochemic theory.

It is not simply the heat that causes distress in a fever patient, but it is the lack of oxygen in the blood due to a deficiency in iron, the carrier of oxygen.

A molecular break in the links of the chain of iron disturbs the continuity of other salts and thus causes more deficiencies. The chloride of potassium (see Lesson V). is usually the first salt called for after the disturbance in iron.

These mighty workers, iron and oxygen, cause all the blood in the body to pass through the heart every three minutes. The lungs contain about one gallon of air at their usual degree of inflation. We breathe, on an average, 1200 breaths per hour; inhale 600 gallons of air per hour and 24,000 gallons daily; and iron and oxygen are the wizards that perform the miracle. When a deficiency in iron occurs, nature—chemical affinity—draws the blood inward from the surface of the body, in order to conserve this life force so that the vital organs, heart, stomach, liver, lungs and brain, may continue to function. But the poor surface circulation allows the pores to close, and thus the waste matter that should escape by this route is turned upon the inner organs, causing exudations, catarrh, pneumonia, pleurisy, etc. But these names are of no consequence. The student will clearly see that iron phosphate is indicated by certain symptoms in whatever part of the organism they may appear. Iron molecules give toughness and strength to the walls of veins and arteries and the minute blood vessels called capillaries (hair-like) and are, therefore, the remedy for hemorrhages.

A child may touch a button that will start a complex machine operating, and yet not understand the science of physics or the mechanism of the machine. So many systems of healing may be the means of starting into action the workmen that may have become dormant. But when the workmen are deficient in the organism, and man's body is a chem-

## Lesson IV

ical formula in operation, it would seem to be the sensible thing to do to furnish the needed chemicals.

There is but one law of chemical operation in vegetable or animal life. When a man understands and co-operates with that operation, he will call into being whatsoever he will; his organism will show forth the glory of omnipresent Spirit and its "fearful and wonderful" mechanism will be the crowning glory of earth.

## LESSON V.

THE Chloride of Potash, or Potassium.
Synonyms—Potassium Chloride, Kali Muriaticum, Kali Chloratum, Kali Chloridum, Potassi Chloridum.

Formula—K Cl.

This salt must not be confused with the chlorate of potash, a poison, chemical formula $K.CLO_3$.

Chloride of potash may be obtained by neutralizing pure aqueous hydrochloric acid with pure potassium carbonate or hydrate.

The cell-salt kali-muriaticum (Potassium chloride) is the mineral worker of the blood that forms fibrin and properly diffuses it through the tissues of the body.

Kali mur molecules are the principal agents used in the chemistry of life to build fibrin into the human organism. The skin that covers the face contains the lines and angles that give expression and thus differentiate one person from another.

In venous blood fibrin amounts to three in one thousand parts; when the molecules of Kali mur fall below the standard in the blood, fibrin thickens, causing what is known as pleurisy, pneumonia, catarrh, diphtheria, etc. When the circulation fails to throw out the thickened fibrin via the glands or mucous membrane, it may stop the action of the heart. Embolus is a Latin word, meaning little lump, or balls; therefore, to die of embolus, or "heart failure" generally means that the heart's action was stopped by little lumps of fibrin clogging the auricles and ventricles of the heart.

When the blood contains the proper amount of kali mur, fibrin is functional and the symptoms referred to above do not manifest.

## Lesson V

Biochemistry has discovered the fact that the cause of embolus, diphtheria, fibroid tumors and fibrinus exudations are not the result of an over-supply of fibrin itself. These symptoms are due to a deficiency in the potash molecules that work with fibrin, diffuse it throughout the organism and build it into tissue.

# LESSON VI.

THE Phosphate of Potash.
    Synonyms—Potassium Phosphate, Kali Phosphoricum, Potassii Phosphas.
    Formula—$K_3PO_4$.

It may be prepared by mixing aqueous phosphoric acid with a sufficient quantity of potash, hydrate or carbonate, until the reaction is slightly alkaline and evaporating. Triturate to 3d or 6th X.

This salt is the great builder of the positive brain cells. Kali phos. unites with albumen and by some subtle alchemy transmutes it and forms gray brain matter.

When the chemical possibilities of this brain builder are fully understood, insane asylums will go out of fashion.

Nervous disorders of all kinds, sleeplessness, paresis, paralysis, irritability, despondency, pessimism, making mountains out of mole hills, crossing broken bridges that do not exist, and borrowing trouble and paying compound interest on the note—all these and many more abnormal conditions that make life a burden are caused by a break in the molecular chain of this nerve and brain builder.

Man has been deficient in understanding because his brain receiver did not vibrate to certain subtle influences; the dynamic cells in gray matter of nerves were not finely attuned and did not respond—hence, sin, or falling short of understanding.

From the teachings of the Chemistry of Life we find that the basis of brain or nerve fluid is a certain mineral salt known as potassium phosphate, or Kali Phos. Kali phosphate is the greatest healing agent known to man, because it is the chemical base of material expression and understanding.

Anything that prevents the formation of new cells as fast as old cells decay or die, disturbs the equilibrium and some

## Lesson VI

pain or other symptom indicates that all is not right. This phenomenon is simply a telegraphic dispatch sent along the nerves to the brain to inform the Ego, the Throne of Understanding, that a deficiency exists—that the material necessary to keep up the processes of life is not sufficiently supplied at a certain place.

And why is it not supplied? Health is that condition of the system where a certain proper degree of heat is maintained, where there is a proper blending of positive and negative electrical influences, and, where every tissue of the body is properly supplied with the right amount of blood, containing all of the elements requisite for building the new cells. This condition can be secured or maintained only by a proper amount of suitable physical exercise, a proper amount of food of a right kind, taken at reasonable intervals, and a judicious adaptation of the clothing to the temperature and occupation.

Anything which breaks up this balance injures, just according to the degree of the adverse influence. In over-eating, the alimentary canal becomes clogged with undigested food, the nutrition, which should be set free to transude through the walls of the intestines to be taken up by the absorbents and carried into the circulation, remains in the fibre of the food and passes out of the body, and, of course, a deficiency at once exists in the blood.

Any disturbance in the molecular motion of these cell-salts in living tissues constitutes disease. This disturbance can be rectified, and the equilibrium re-established by administering a small dose of the same mineral salts in molecular form.

The Biochemic System of Medicine is founded on physiology, anatomy, cellular pathology and chemistry, as set forth by Schuessler, Huxley, Tyndall, Virchow, Liebig, Valentin, Goullon, Moleschott and Walker, of Europe, and many noted scientists of our own land. Professor Moleschott, of the University of Rome, says in his work, The Circulation of Life: "The structure and vital power of the organs are conditional upon the necessary quantity of the inorganic salts of the blood."

Dr. Schuessler says that these words awakened in him the idea of employing for healing purposes the inorganic salts alone. Schuessler says in his Therapeutics: "My system or method of procedure is direct Biochemistry, because I use only tissue cell-salts, substances which are homogeneous to those contained in the diseased tissue. These salts, used properly, in a proper potency, cure all curable disease."

# LESSON VII.

SULPHATE of Potash.
Synonyms—Potassium Sulphate, Kali Sulphos, Potassae Sulphos, Kali Sulphate.
Formula—$K_2SO_4$.

The microscope reveals the fact that, when the body is in health, little jets of steam are constantly escaping from the seven million pores of the skin. The human body is a furnace and steam engine. The stomach and bowels burn food by chemical operation as truly as the furnace of a locomotive consumes by combustion. In the case of the locomotive the burning of coal furnishes force which vibrates water and causes an expansion (rate of motion) that we name steam.

The average area of skin is estimated to be about 2000 square inches. The atmospheric pressure, being fourteen pounds to the square inch, a person of medium size is subject to a pressure of 28,000 pounds.

Each square inch of skin contains 3500 sweat tubes, or perspiratory pores( each of which may be likened to a little drain tile) one-fourth of an inch in length, an aggregate length of the entire surface of the body of 201,166 feet, or a tile for draining the body nearly forty miles in length.

Let me repeat, for it is very important, the stomach is the furnace of man's body, and, by the process of digestion, burns up food and furnishes the force to run the human engine, and thus enable it to inhale air, the material for blood, as water is the material for steam. In the manufacture of blood, through the complex operation of air passing through lung-cells, arteries, etc., a certain amount of water is changed to steam, a portion of which must escape through the safety valves provided by Divine Intelligence for that purpose. Sometimes the pores become clogged, and prevent the steam from escaping; then the vibration of the body changes and the person is sick. In many cases a disturbance in oil is the cause of the trouble. Potassium sulphate has an affinity for oil; it is the maker and distributor of oil. When this salt falls below the standard in quantity, in the human organism, oil becomes non-functional—too thick, and thus clogs the pores.

And does it not seem strange that medical science, that boasts of such great progress, can invent no better term than "bad color" for these chemical results?

Kali sulph. is found in considerable quantities in scalp and hair. When this salt falls below the standard, dandruff or eruptions, secreting yellowish, thin, oily matter, or falling out of hair, is the result.

Kali sulph. is a wonderful salt, and its operation in the divine laboratory of man's body, where it manufactures oil, is the miracle of the chemistry of life.

Oil is made by the union of the sulphate of potassium (Potash) with albuminoids and aerial elements.

A deficiency of sulphate of potash in the molecules is the cause of oily, slimy, yellowish exudations from any orifice of the body, or from any glandular swellings, abscesses, cancers, etc.

# LESSON VIII.

PHOSPHATE of Magnesia.
Synonyms—Magnesium Phosphorica.
Formula—$Mg_3 PO_4{}_2$.

This cell-salt may be made by mixing Phosphate of Soda with Sulphate of Magnesia. This salt is found chiefly in the white fibres of nerves and muscles. The tissues of nerves and muscles are composed of many very fine threads or strands of different colors, each acting as a special telegraph wire, each one having a certain conductive power or quality, i. e., special chemical affinity—for certain organic substances, oil or albumen, through and by which the organism is materialized and the process or operations of life are carried on. The imagination might easily conceive the idea that these delicate infinitesimal fibres are strings of the Human Harp, and that molecular minerals are the fingers of infinite Eneregy, striking notes of some Divine Anthem.

The white fibres of nerves and muscles need the dynamic action of Magnesia Phosphate, especially to keep them in proper tune, or function, for, by its chemical action on albumen, the special fluid for white nerve or muscle fibre is formed. When the supply of this salt falls below the standard, cramps, sharp shooting pains, or some spasmodic condition, prevails. Such symptoms are simply calls of nature for more magnesia.

The human body is composed of perfect principles, gases, minerals, molecules or atoms; but these builders of flesh and bone are not always properly adjusted.

The planks or bricks used in building houses may be endlessly diversified in arrangement and yet be perfect material. Therefore, we must conclude that symptoms of disease are dispatches sent to the brain—the throne of understanding—calling for the worker, the builder, needed to carry on Life's

## The Chemistry and Wonders of the Human Body

work in flesh. Cases of Chorea (St. Vitus' dance) are cured by the proper use of magnesia phosphate. This salt is the great remedy for nearly all heart troubles, except embolus. (See Kali Mur., Lesson V.)

The white fibres in the delicate strands that compose the tissue of nerves are controlled by the molecules of magnesia; when these workers are deficient in amount, these live wires contract or cramp up in knots, of course, infinitely small, the effect of which is a sharp shooting pain, as in so-called neuralgia, or sciatica. The word neuralgia is from (a) Latin for nerves (b) Greek for pain, and therefore simply means nerve pain. English expresses the effect quite as well as other languages. But the idea generally prevails that neuralgia means a thing unknown and undefinable that causes the pain, and that the name of this unknown thing is neuralgia.

The pain is simply a dispatch, or words, asking for magnesia phosphate,

A deficiency of this salt in the muscular and nerve tissue of the walls of the stomach causes contraction, cramps, which reduces the cavity of the stomach. In order to meet this condition and prevent a collapse, such as is formed by natural chemical process from material at hand and by expansion, magnesium phosphate produces a counter-force that wards off more serious results. Magnesium phosphate relieves such conditions immediately, thus demonstrating the theory that to supply the deficient tissue-builder is the natural method of cure. The phosphate of lime often supplements magnesia. (See page 95, "The Biochemic System of Medicine.")

This wonderful salt is the true antispasmodic remedy. It has cured cases of chorea, or St. Vitus' dance, in from two to four weeks. For all heart troubles caused by distension of the cardiac portion of the stomach, thus interfering with the action of the heart, it is the soverign remedy.

Dr. Baericke, one of the leading homeopathic physicians of the Pacific Coast, says: "Magnesia phosphate is a magnificent remedy in all spasmodic diseases."

It is plainly evident that the wonderful fluids of the human body are manufactured in the chemical laboratory of the

## The Chemistry of Wisdom

organism. The particles of magnesia evidently contain within themselves the power and potency to create the white fibre nerve fluid by using albuminous substances as a basis, and then calling to its aid the spirit of life, oxygen. Each one of the inorganic salts knows how to make some fluid or tissue of the human machine. "And the tree bore twelve manner of fruits and its leaves were for the healing of the nations."

Like the phosphate of potash (see Lesson VI), magnesia phosphate is a nerve and brain salt, and, when we consider the wonders of the brain and its marvelous mechanism, we must recognize the great importance of the wizard workmen which labor for three score and ten years without an instant rest.

# LESSON IX.

SODIUM Chloride.
Synonyms—Natrum muriaticum, Sodii Chloridium, Chloruretum of Sodicum, Common table salt.
Formula—Na Cl.

A combination of sodium and chlorine forms the mineral known as common salt. This mineral absorbs water. The circulation or distribution of water in the human organism is due to the chemical action of the molecules of sodium chloride.

This inorganic cell-salt is the bearer and distributor of water.

Sodium chloride must be triturated up to the 3d or 6th biochemic potency before using it as an agent to supply deficiencies in the water carrier molecules arising from a lack of this salt.

Water constitutes over 70 per cent of the human body, therefore, the carriers of water must be in like proportion. There is more sodium chloride in the ashes of a cremated body than any of the 12 mineral salts, except the phosphate of lime, which composes 57 per cent of bone structure.

Through its affinity for water, this salt assists in carrying on the process of life in the human organism as well as in all vegetable tissue.

When there is a deficiency in the molecules of the water bearer, salt, the molecular continuity of water is broken, and, as a result, too much water will appear at a certain point and corresponding dryness or lack of water at other places. Example: Watery discharge from nasal passages and constipation of the bowels.

Sunstroke and delirium tremens are caused by a break in the supply of this salt, which causes water to press or crowd the membranes of the cerebellum (lower brain), and thus prevent the magnetic vibrations from the cerebrum (upper brain) from passing to the solar plexus, or central brain.

## Lesson IX

Crude soda cannot be taken up by mucous membrane absorbents and carried into the circulation. The sodium molecules found in the blood have been received from vegetable tissue which drew these salts from the soil in high potency. The mineral, or cell-salts, can also be prepared and are prepared in biochemic or homeopathic potency as the trituration of Nature's laboratory in the physiology of plant growth, and then, thoroughly mixed with sugar of milk and pressed into tablets ready to be taken internally, supply deficiencies in the human organism. A lack of the proper amount of these basic mineral salts, twelve in number, is the cause of all so-called disease.

Common table salt does not enter the blood, being too coarse to enter the delicate tubes of mucous membrane absorbent, but this salt does distribute water along the intestinal tract.

Professor Leibig says in his Chemical Letters, that muriatic acid, when diluted a thousand fold with water, dissolves, with ease, at the temperature of the body, fibrin and gluten, and this solvent power does not increase, but diminishes, if the proportion of acid in the dilution be increased.

Air contains 78 per cent of nitrogen gas, believed by scientists to be mineral in ultimate potency. Minerals are formed by the precipitation of nitrogen gas. Differentation is attained by the proportion of oxygen and aqueous vapor (hydrogen) that unites with nitrogen.

A deficiency in sodium chloride causes the water in the blood serum to become inert and non-functional. By its affinity for water, sodium chloride assists in the biological operation in blood and tissue.

As will be seen in Lesson XI, sodium sulphate eliminates an excess of water in the blood and thus regulates the supply; while sodium chloride properly distributes water in the physiology of animal or vegetable forms.

# LESSON X.

PHOSPHATE of Soda.
Synonyms—Natrum phosphate, Sodium phosphate, Natri Phosphate, Phosphos Natricus, Sodae Phosphate.

Formula—$Na_2HPO_4; _{12} H_2O$; $Na_2 PO_4X_{12}H_3O$.

This alkaline cell-salt is made from bone ash or by neutralizing orthophosphoric acid with carbonate of sodium.

Sodium, or natrum, phosphate holds the balance between acids and normal fluids of the human body.

Acid is organic and can be chemically split into two or more elements, thus destroying the formula that makes the chemical rate of motion called acid.

Acid conditions are not due to an excess of acid in the blood, bile or gastric fluids. Supply the alkaline salt sodium phosphate, and acid will chemically change to normal fluids.

A certain amount of acid is necessary, and always present in the blood, nerve, stomach and liver fluids. The apparent excess of acid is nearly always due to a deficiency in the alkaline, salt.

Acid, in alchemical lore, is represented as Satan, Saturn, while sodium phosphate symbolizes Christ, Venus. An absence of the Christ principle gives license to Satan to run riot in the Holy Temple. The advent of Christ drives the exile out with a whip of thongs. Reference to the temple, in the figurative language of the Bible and New Testament, always symbolizes the human organism. "Know ye not that your bodies are the Temple of the living God?"

Solomon's temple is an allegory of the physical body of man and woman. Soul—of man's temple—the house, church, Beth or temple made without sound of "saw or hammer."

Hate, envy, criticism, jealousy, competition, selfishness, war, suicide and murder are largely caused by acid condi-

## Lesson X

tions of the blood, phoducing changes by chemical poison and irritation of the brain cells, the keys upon which Soul plays "Divine Harmonies" or plays "fantastic tricks before high Heaven," according to the arrangement of chemical molecules in the wondrous laboratory of the soul.

Without a proper balance of the alkaline salt, the agent of peace and love, man is fit for "treason, stratagem and spoils."

The chemistry of life points to the reason why man is unbalanced. Poise of tissue, nerve fluid, blood and brain cells, as well as muscular fibre, are conditions precedent to mental or soul poise. A man thinks and acts according to the organism in or through which the Ego operates.

The basis of the human body are twelve minerals, lime, iron, potash, silica, sodium, magnesia, etc., etc. They are found in the ashes of a cremated body, or, in reality, constitute the ashes. Before man ceases to be sick, before envy, strife, hatred, competition, selfishness, war and murder cease on earth, man must build his body on a plan that will express mind on the plane of altruism and love.

The perfectly balanced body will enable the mind to cognize oneness of being. From this concept comes peace on earth and good-will between man and man.

The knowledge of Life Chemistry will bring man to his Divine estate in the Kingdom of Harmony and Love. This Kingdom is forming in the chemicalizing mass of God's creative compounds. Out from the chemistry of elements, principles, monads and molecules; out of oxygen, hydrogen, nitrogen, carbon, helium, uranium, radium, aurium, argentum, sodium, potassium and iron—out from molecules composing the body of universal energy, a man and woman will be born—real Sons of God—who will bear away the sins of the world.

# LESSON XI.

**S**ULPHATE of Soda.
Synonyms—Natrum Sulphate, Sodium Sulphuricum, Sodae or Sodii Sulphas, Glauber's Salts.
Formula—$Na_2 SO_4\ 10\ H_2O$.

This may be obtained by the action of Sulphuric acid on sodium chloride (common salt).

This cell-salt is found in the intercellular fluids, liver and pancreas. Its principal work is to regulate the supply of water in the human organism.

The blood becomes overcharged with water, either from the oxidation of organic matter or from inhaling air that contains more aqueous vapor (water) than is required to produce normal blood. This condition of air is liable to prevail whenever the temperature is above 70 degrees.

One molecule of nat. sulph. has the power, chemical intelligence, to take up and carry away two molecules, or twice its bulk, of water. The blood does not become overcharged with water from water taken into the stomach, but from the water lifted by expansion caused by heat above 70 degrees and held in the air and thus breathed into the arteries through the lungs. By the above we see that there is more work for this salt in hot weather than during cold weather. So-called malaria, Latin for bad air, is due to a lack of this tissue salt. Water, lifted from swamps or clear streams or lakes by the action of the sun's heat, is the same, for heat does not evaporate and lift poisonous, disintegrating organic matter from a swamp or marsh, but the water only.

Therefore, it is not some impurity in the air that causes chills, etc., but an oversupply of water which thins the bile and distributes it through the organism. Nature's effort to get rid of the surplus water by nervous, muscular and vascular contraction, on the principle of wringing water from a cloth, causes the spasm called chills. Proof of this theory

## Lesson XI

is found in the fact that perspiration follows the chill. It generally requires about forty-eight hours to again overcharge the blood and bring on another chill.

Cold, dry air always cures chill, and be it known that all cold air is dry air. The cure for chills when cold air cannot be had is sodium sulphate in biochemic potency. Yellow fever is caused by too much water in bile and other liver fluids. These fluids are distributed through the system, and in their union with oil, albumen, etc., become vitiated and cause the yellow skin.

Sodium Sulph. in crude form, is known as Glauber's Salts, and is too coarse to be taken up by the mucous membrane absorbents and carried into the circulation; it must be triturated with sugar of milk, according to the biochemic method, up to the 3d or 6th decimal before using a remedy to supply the blood. Glauber's Salts, crude sodium sulphate, acts as a cathartic, and cathartics are never used in the biochemic system of healing.

When man learns to keep his blood at the proper rate of motion by the proper dynamos—the mineral salts—he will not fear fevers, microbes, mosquitoes, nor devils.

# LESSON XII.

## SILICA.

Synonyms—Silica, silic, oxide, white pebble or common quartz. Chemical abbreviation, Si.

Made by fusing crude silica with carbonate of soda; dissolve the residue, filter and precipitate by hydrochloride acid.

This product must be triturated as per biochemic process before using internally.

This salt is the surgeon of the human organism. Silica is found in hair, skin, nails, periosteum, the membrane covering and protecting bone, the nerve sheath, called neurilemma, and a trace is found in bone tissue. The surgical qualities of silica lie in the fact that its particles are sharp cornered. A piece of quartz is a sample of the finer particles. Reduce silica to an impalpable powder and the microscope reveals the fact that the molecules are still pointed and jagged like a large piece of quartz rock. In all cases where it becomes necessary that decaying organic matter be discharged from any part of the body by the process of suppuration, these sharp pointed particles are pushed by the marvelous intelligence which operates without ceasing, day and night, in the wondrous human Beth, and like a lancet cuts a passage to the surface for the discharge of pus. Nowhere in all the records of physiology or biological research can anything be found more wonderful than the chemical and mechanical operation of this Divine artisan.

The bone covering is made strong and firm by silica. In case of boils or anthrax carbuncle, the biochemist loses no time searching for "anthrax bacilli," or germs, nor does he experiment with imaginary germ-killing serum, but simply furnishes nature with tools with which the necessary work may be accomplished.

Silica gives the glossy finish to hair and nails. A stalk of corn or straw of wheat, oats or barley would not stand upright, except they contained this mineral.

# RELATION OF THE MINERAL SALTS OF THE BODY TO THE SIGNS OF THE ZODIAC

TO those who object to linking chemistry with astrology, the writer has this to say:

The Cosmic Law is not in the least disturbed by negative statements of the ignorant individual. Those investigators of natural phenomena, who delve deeply to find Truth, pay little heed to the babbler who says, "I can't understand how the zodiacal signs can have any relation to the cell-salts of the human body." The sole reason that he "cannot understand" is because he never tried to understand.

A little earnest, patient study will open the understanding of any one possessed of ordinary intelligence and make plain the great truth that the UNIverse is what the word implies, i. e., one verse.

It logically follows that all parts of one thing are susceptible to the operation of any part.

The human body is an epitome of the cosmos.

Each sign of the Zodiac is represented by the twelve functions of the body and the position of the Sun at birth.

Therefore the cell-salt corresponding with the Sign of the Zodiac and function of the body is consumed more rapidly than other salts and needs an extra amount to supply the deficiency caused by the Sun's influence at that particular time.

Space will only permit a brief statement of the awakening of humanity to great occult truths. However, the following from India will indicate the trend of new thought: "Dr. Carey's remarkable researches in the domain of healing art have left no stone unturned. His discovery of the Zodiacal cell-salts has added a new page in the genesis of healing art," writes Swaminatha Bomiah, M.B., Ph.D., No. 105 Armenian street, Madras, India.

# THE TWELVE CELL SALTS OF THE ZODIAC
## ARIES: "THE LAMB OF GOD"
### March 21 to April 19

ASTROLOGERS have for many years waited for the coming discovery of a planet to rule the head or brain of man, symbolized in the "Grand Man" of the heavens by the celestial sign of the zodiac, regnant from March 21st to April 19. This sign is known as Aries—the Ram or Lamb.

In ancient lore Aries was known as the "Lamb of Gad," or God, which represents the head or brain. The brain controls and directs the body and mind of man. The brain itself, however, is a receiver operated upon by celestial influences or angles (angels) and must operate according to the directing force or intelligence of its source of power.

Man has been deficient in understanding because his brain receiver did not vibrate to certain subtle influences. The dynamic cells in the gray matter of the nerves were not finely attuned and did not respond—hence sin, or falling short of understanding.

From the teachings of the Chemistry of Life we find that the basis of the brain or nerve fluid is a certain mineral salt known as potassium phosphate, or Kali Phos.

A deficiency in this brain constituent means "sin," or a falling short of judgment or proper comprehension. With the advent of the Aries Lord, ruler, or planet, cell-salts are rapidly coming to the fore as the basis of material healing. Kali phosphate is a great healing agent because it is the chemical base of material expression and understanding.

The cell-salts of the human organism are now being prepared for use, while poisonous drugs are being discarded everywhere. Kali phosphate is the especial birth salt for those born between March 21 and April 19.

## The Twelve Salts of the Zodiac

These people are brain workers, earnest, executive and determined—thus do they rapidly use up the brain vitalizers.

The Aries gems are amethyst and diamond.

The astral colors are white and rose pink.

In Bible alchemy Aries represents Gad, the seventh son of Jacob, and means "armed and prepared"—thus it is said when in trouble or danger, "keep your head."

In the symbolism of the New Testament, Aries corresponds with the disciple Thomas. Aries people are natural doubters until they figure a thing out for themselves.

# TAURUS—THE "WINGED BULL" OF THE ZODIAC
## April 19 to May 20

THE ancients were not "primitive men." There never was a first man, nor a primitive man. Man is an eternal verity—the Truth, and Truth never had a beginning.

The Winged Bull of Nineveh is a symbol of the great truth that substance is materialized air, and that all so-called solid substances may be resolved into air.

Taurus is an earth sign, but earth is precipitated aerial elements. This chemical fact was known to the scientists of the Taurian age (over 4000 years ago); therefore they carved the emblem of their Zodiacal sign with wings.

Those born between the dates, April 19th and May 20th, can descend very deep into materiality or soar "High as that Heaven where Taurus wheels," as written by Edwin Markham, who is a Taurus native.

What can be finer than the following from this noted Taurian, he who has sprouted the wings of spiritual concept:

"It is a vision waiting and aware,
　And you must bring it down, oh, men of worth,
　Bring down the New Republic hung in air
　And make for it foundations on the Earth."

*Air is the "raw material" for blood,* and when it is drawn in, or breathed in, rather, by the "Infinite Alchemist," to the blood vessels, it unites with the philosopher's stone, mineral salts, and in the human laboratory creates blood.

So, then, blood is the elixir of life, the 'Ichor of the Gods."

The sulphate of sodium, known to druggists as Nat. Sulph., chemically corresponds to the physical and mental characteristics of those born in the Taurus month.

## Man's Divine Estate

Taurus is represented by the cerebellum, or lower brain, and neck.

A deficiency in Nat. Sulph. in the blood is always manifested by pains in the back of head, sometimes extending down the spine, and then affecting the liver.

The first cell-salt to become deficient in symptoms of disease in the Taurus native is Nat. Sulph.

The chief office of Nat. Sulph. is to eliminate the excess of water from the body.

In hot weather the atmosphere becomes heavily laden with water and is thus breathed into the blood through the lungs.

One molecule of the Taurus salt has the chemical power to take up and carry out of the system two molecules of water.

Blood does not become overcharged with water from the water we drink, but from an atmosphere overcharged with aqueous vapor drawn from water in rivers, lakes or swamps, by heat of the sun above 70 degrees in shade.

The more surplus water there is to be thrown out of blood, the more sodium sulphate is required.

All so-called bilious or malarial troubles are simply a chemical effect or action caused by deficient sulphate of soda. (See Article on Intermittent Fever, page 87.)

Chills and fever are Nature's method of getting rid of surplus water by squeezing it out of the blood through violent muscular, nervous and vascular spasms.

No "shakes" or ague can occur if blood be properly balanced chemically.

Governing planet: Venus.

Gems: Moss-agate and emerald.

Astral colors: Red and lemon yellow.

In Bible Alchemy Taurus represents Asher, the eighth son of Jacob, and means blessedness or happiness.

In the symbolism of the New Testament, Taurus corresponds with the disciple Thaddeus, meaning firmness, or led by love.

# THE CHEMISTRY OF GEMINI
## May 20 to June 21

ONE of the chief characteristics of the Gemini Native is expression. The cell-salt kali muriaticum (potassium chloride) is the mineral worker of blood that forms fibrine and properly diffuses it throughout the tissues of the body.

This salt must not be confused with the chlorate of potash, a poison (chemical formulae $K.ClO_3$).

The formulae of the chloride of potassium (kali mur) is $K.Cl$.

Kali mur molecules are the principal agents used in the chemistry of life to build fibrine into the human organism. The skin that covers the face contains the lines and angles that give expression and thus differentiate one person from another; therefore the maker of fibrine has been designated as the birth salt of the Gemini native.

In venous blood fibrine amounts to three in one thousand parts. When the molecules of kali mur falls below the standard, the blood fibrine thickens, causing what is known as pleurisy pneumonia, catarrh, diphtheria, etc. When the circulation fails to throw out the thickened fibrine via the glands or mucus membrane, it may stop the action of the heart. Embolus is a Latin word meaning little lump, or balls; therefore to die of embolus, or "heart failure," generally means that the heart's action was stopped by little lumps of fibrine clogging the auricles and ventricles of the heart.

When the blood contains the proper amount of kali mur, fibrine is functional and the symptoms referred to above do not manifest. Gemini means twins. Gemini is the sign which governs the United States.

The astral colors of Gemini are red, white and blue. While those who made our first flag and chose the colors

## The Chemistry of Gemini

personally knew nothing of astrology, yet the Cosmic Law worked its will to give America the "red, white and blue."

Mercury is the governing planet of Gemini. The gems are beryl, aquamarine and dark blue stones. In Bible alchemy Gemini represents Issachar, the ninth son of Jacob, and means price, reward or recompense. In the symbolism, allegories of the New Testament, Gemini corresponds with the disciple Judas, which means service or necessity. The perverted ideas of an ignorant dark-age priesthood made "service and necessity" infamous by a literal rendering of the alchemical symbol, but during the present aquarian age, the Judas symbol will be understood and the disciple of "service" will no longer have to submit to "third degree methods."

> "Each life is fed
> From many a fountain-head,
> Tides that we never know
> Into our being flow,
> And rays from the remotest star
> Converge to made us what we are."

# CANCER: THE CHEMISTRY OF THE "CRAB"
### June 21 to July 22

CANCER is the Mother Sign of the Zodiac.

The mother's breast is the ego's first home after taking on flesh and "rending the Veil of Isis."

The tenacity of those born between the dates, June 21 and July 22, in holding onto a home or dwelling place is well illustrated by the crab's grip, and, also, by the fact that it carries its house along wherever it goes in order that it may be sure of a dwelling.

The Angles (Angels) of the twelve Zodiacal Signs materialize their vitalities in the human microcosm. Through the operation of chemistry, energy creating, the intelligent molecules of Divine Substance make the "Word flesh."

The corner stone in the chemistry of the crab is the inorganic salt fluoride of lime, known in pharmacy language as Calcarea Flurica. It is a combination of fluorine and lime.

When this cell-salt is deficient in the blood, physical and mental disease (not-at-ease) is the result. Elastic fiber is formed by the union of the fluoride of lime with albuminoids, whether in the rubber tree or the human body. All relaxed conditions of tissue (varicose veins and kindred ailments) are due to a lack of sufficient amount of elastic fiber to "rubber" the tissue and hold it in place.

When elastic fiber is deficient in tissue of membrane between the upper and lower brain poles—cerebrum and cerebellum—there results a "sagging apart" of the positive and negative poles of the dynamo that runs the machinery of man.

An unfailing sign or symptom of this deficiency is a groundless fear of financial ruin.

## Cancer: The Chemistry of the "Crab"

While those born in any of the twelve signs may sometimes be deficient in Cal. Fluo., due to Mars or Mercury (or both) in Cancer at birth, Cancer people are more liable to symptoms, indicating a lack of this elastic fiber-builder than are those born in other signs.

Why should we search Latin and Greek lexicons to find a name for the result of a deficiency in some of the mineral constituents of blood? If we find a briar in our flesh, we say so in the plainest speech; we do not say, "I have got the briatitis or splintraligia."

When we know that a deficiency in the cell-salts of the blood causes the symptoms that medical ignorance has dignified and personified with names that nobody knows the meaning of, we will know how to scientifically heal by the unalterable law of the chemistry of life. When we learn the cause of disease, then, and not before, will we prevent disease.

Not through quarantine, nor disinfectants, nor "Boards of Health" will man reach the long sought plane of health; not through affirmations of health, nor denials of disease will bodily regeneration be wrought; not by dieting or fasting or "Fletcherizing" or suggesting, will the elixir of life and the philosopher's stone be found.

The "mercury of the sages" and the "Hidden manna" are not constituents of health foods.

Victims of salt baths and massage are bald before their time, and the alcohol, steam and Turkish bath fiends die young.

"Sic transit gloria mundi."

When a man's body is made chemically perfect, the operations of mind will perfectly express.

Gems belonging to the sign of the breast are black onyx and emerald; astral colors, green and russet, brown.

Cancer is represented by Zebulum, the tenth son of Jacob, and means dwelling place or habitation.

Matthew is the Cancer disciple.

# LEO: THE HEART OF THE ZODIAC
## July 22 to August 22

THE Sun overflows with divine energy. It is the "brewpot" that forever filters and scatters the "Elixir of Life."

Those born while the Sun is passing through Leo, July 22 to August 22, receive the heart vibrations, or pulses, of the Grand Man, or "Circle of Beasts." All the blood in the body passes through the heart and the Leo native is the recipient of every quality and possibility contained in the great "Alchemical Vase," the "Son of Heaven."

The impulsive traits of Leo people are symboled in the pulse which is a reflex of heart throbs.

The astronomer, by the unerring law of mathematics applied to space, proportion, and the so-far-discovered wheels and cogs of the uni-machine, can tell where a certain planet must be located before the telescope has verified the prediction. So the astro-biochemist knows there must of necessity be a blood mineral and tissue builder to correspond with the materialized angle (angel) of the circle of the Zodiac.

The phosphate of magnesia, in biochemic therapeutics, is the remedy for all spasmodic impulsive symptoms. This salt supplies the deficient worker or builder in such cases and thus restores normal conditions. A lack of muscular force, or nerve vigor, indicates a disturbance in the operation of the heart cell-salt, magnesia phosphate, which gives the "Lion's spring," or impulse, to the blood that throbs through the heart.

Leo is ruled by the Sun, and the children of that celestial sign are natural sun worshipers.

Gold must contain a small percent of alloy or base metal before it can be used commercially. Likewise the "Gold of Ophir"—Sun's rays, or vibration—must contain a high

## Leo: The Heart of the Zodiac

potency of the earth salt, magnesia, in order to be available for use in bodily function. Thus through the chemical action of the inorganic (mineral and water) in the organic, Sun's rays and ether, does the volatile become fixed, and the word becomes flesh.

Leo people consume their birth salts more rapidly than they consume any of the other salts of the blood; hence are often deficient in magnesium. Crude magnesia is too coarse to enter the blood through the delicate mucus membrane absorbents, and must be prepared according to the biochemic method before taken to supply the blood.

Gems of Leo are ruby and diamond.

Astral colors, red and green.

The eleventh child of Jacob, Dinah, represents Leo and means judged. Simon is the Leo disciple.

# VIRGO: THE VIRGIN MARY
## August 22 to September 23

VIRGIN means pure. Mary, Marie, or Mare (Mar) means water. The letter M is simply the sign of Aquarius, "The Water Bearer."

Virgin Mary means pure sea, or water.

Jesus is derived from a Greek word, meaning fish. Out of the pure sea, or water, comes fish. Out of woman's body comes the "word made flesh." All substance comes forth from air, which is a higher potency of water.

All substance is fish, or the substance of Jesus.

This substance is made to say, "Eat, this is My body; drink, this is My blood."

There is nothing from which flesh and blood can be made, but the one universal Air, Energy, or Spirit, in which man has his being.

All tangible elements are the effects of certain rates of motion of the intangible and unseen elements. Nitrogen gas is mineral in solution, or ultimate potency.

Oil is made by the union of the sulphate of potassium (potash), with albuminoids and aerial elements.

The first element that is disturbed in the organism of those born in the celestial sign Virgo is oil; this break in the function of oil shows a deficiency in potassium sulphate, known in pharmacy as kali sulph.

Virgo is represented in the human body by the stomach and bowels, the laboratory in which food is consumed as fuel to set free the minerals, in order that they may enter the blood through the mucous membrane absorbents.

## Virgo: The Virgin Mary

The letter X in Hebrew is Samech or Stomach. X, or cross, means crucifixion, or change-transmutation.

Virgo people are discriminating, analytical and critical.

The microscope reveals the fact that when the body is in health little jets of steam are constantly escaping from the seven million pores of the skin. A deficiency in kali sulph. molecules causes the oil in the tissue to thicken and clog these safety valves of the human engine, thus turning heat and secretions back upon the inner organs, lungs, pleura, membrane of nasal passages, etc. And does it not seem strange that medical science, that boasts of such great progress, can invent no better term than "bad cold" for these chemical results?

Kali sulph. is found in considerable quantities in the scalp and hair. When this salt falls below the standard, dandruff or eruptions, secreting yellowish thin, oily matter or falling out of the hair, is the result.

Kali sulph. is a wonderful salt, and its operation in the divine laboratory of man's body, where it manufactures oil, is the miracle of the chemistry of life.

Governing planet, Mercury. Gems, pink jasper and hyacinth. Astral colors, gold and black.

In Bible alchemy Virgo is represented by Joseph, the twelfth son of Jacob, and means: To increase power, or "son of the right hand."

Virgo corresponds with the disciple Bartholomew.

The thirteenth child of Jacob, the Circle, is Benjamin. See 35th Chapter of Genesis. Also explained in the great book, "God-man, the Word Made Flesh."

# LIBRA: THE LOINS
## SEPTEMBER 23 TO OCTOBER 23

THIS alkaline cell-salt is made from bone ash or by neutralizing orthophosphoric acid with carbonate of sodium.

Libra is a Latin word, meaning scales or balance. Sodium, or natrum, phosphate holds the balance between acids and the normal fluids of the human body.

Acid is organic and can be chemically split into two or more elements, thus destroying the formula that makes the chemical rate of motion called acid.

A certain amount of acid is necessary, and is always present in the blood, nerves, stomach and liver fluids. The apparent excess of acid is nearly always due to a deficiency in the alkaline, Libra, salt.

Acid, in alchemical lore, is represented as Satan (Saturn), while sodium phosphate symbols Christ (Venus). An absence of the Christ principle gives license to Satan to run riot in the Holy Temple. The advent of Christ drives the evil out with a whip of thongs. Reference to the temple in the figurative language of the Bible and New Testament always symbols the human organism. "Know ye not that your bodies are the Temple of the living God?"

Solomon's temple is an allegory of the physical body of man and woman. Soul-of-man's-temple—the house, church, Beth or temple made without sound of "saw or hammer."

The thirty-third vertebrae of spinal column represents the thirty-third degree of Masonry.

Hate, envy, criticism, jealousy, competition, selfishness, war, suicide and murder are largely caused by acid conditions of the blood, producing changes by chemical poisons and irritation of the brain cells, the keys upon which Soul plays "Divine Harmonies" or plays "fantastic tricks before

## Libra: The Loins

high heaven," according to the arrangement of chemical molecules in the wondrous laboratory of the soul.

Without a proper balance of the Venus salt, the agent of peace and love, man is fit for "treason, strategems and spoils."

The people of the world never needed the alkaline or Libra salt more than they do at the present time, while wars and rumors of wars strut upon the Stage of Life (1918).

The Sun enters Libra September 23 and remains until October 23.

Governing planet—Venus.

Gems—Diamond and opal.

Astral colors are black, crimson and light blue.

Libra is an air sign.

In Bible alchemy Libra represents Reuben, the first son of Jacob. Reuben means Vision of the Sun.

In the symbolism of the New Testament, Libra corresponds with the disciple Peter.

Peter is derived from Petra, a stone or mineral.

On thee, Peter (mineral), will I build my church, viz., beth, house, body or temple.

# INFLUENCE OF SUN ON VIBRATION OF BLOOD AT BIRTH
## SCORPIO—OCTOBER 24 TO NOVEMBER 22

FROM Scorpion to "White Eagle" may seem a very long journey to one who has not learned the science of patience or realized that time is an illusion of the physical senses.

The Zodiacal sign Scorpio is represented in human material organism by the sexual functions.

The esoteric meaning of sex is based in mathematics, the body being a mathematical fact. Sex in Sancrit means Six.

"Six days of Creation" simply means that all creation, or formation, from self-existing substance, is by and through the operation of sex principle—the only principle.

Three means male, father, the spirit of the male, and son; this trinity forms or constitutes one pole of Being, Energy or Life—the positive pole.

The negative pole, female trinity; female spirit or mother and daughter.

Thus two threes or trinities produce six or sex, the operation of which is the cause of all manifestation. Those who understand fully realize the truth of the New Testament statement, "There is no other name under heaven whereby ye may be saved (materialized and sustained), except through Jesus Christ and Him Crucified." By tracing the words Jesus and Crucify (also Christ) to their roots a wonderful world of truth appears to the understanding.*

The possibilities of Scorpio people are boundless after they have passed through trials and tribulations, viz.: Crucifixion or crossification.

One of the cell-salts of the blood, calcarea, sulphate, is the mineral ("stone") that especially corresponds to the Scorpio nature. Crude Calcarea sulphate is gypsum or sulphate of lime.

---

*Explained in God-man: The word made Flesh.

*Influences of the Sun*

While in crude form lime is of little value, but add water and thus transmute it by changing its chemical formation, and plaster of paris is formed, a substance useful and ornamental. Every person, born between October 24 and November 22, should well consider this wonderful alchemical operation of their esoteric stone and thus realize the possibilities in store for them on their journey to the "Eyric of the White Eagle."

Scorpio people are natural magnetic healers, especially after having passed through the waters of adversity, as heat is caused by the union of water and lime.

Scorpio is a water sign, governed by Mars. Mars is "a doer of things," also fiery at times, therefore, it is well that the Scorpio native take heed les he sometimes "boil over."

In Bible alchemy, Scorpio is represented by Simeon, the second son of Jacob. Simeon means "hears and obeys." In the symbolism of the New Testament Scorpio corresponds with the disciple Andrew, to create or ascend.

The gems are topaz and malachite; astral colors, golden brown and black.

A break in the molecular chain of the Scorpio salt, caused by a deficiency of that material in the blood, is the primal cause of all the so-called diseases of these people. This disturbances, not only causes symptoms called disease in physical functions, but it disturbs the astral fluids and gray matter of brain cells and thereby changes the operation of mind into inharmony. Sin means to lack or fall short; thus chemical deficiencies in life's chemistry cause sin.

When man learns to supply his body with the proper dynamics, he will "wash away his sins with the blood of Christ"—blood made with the "White Stone." Christ is Greek for oil.

Calcium sulphate should not be taken internally in crude form; in order to be taken up by absorbents of mucus membrane the lime salt must be triturated, according to the Biochemic method, up to 3rd or 6th. By this method lime may be rendered as fine as the molecules contained in grain, fruit or vegetables.

## The Chemistry and Wonders of the Human Body

Blood contains three forms of lime. Lime and fluorine for Cancer sign; lime and phosphorus for Capricorn sign, and lime and sulphur for Scorpio.

Lime should never be used internally below 3rd decimal trituration.

# THE CHEMISTRY OF SAGITTARIUS
## November 22 to December 22

THE mineral or cell-salt of the blood corresponding to Sagittarius is Silica.

Synonyms: Silicea, silici oxide, white pebble or common quartz. Chemical abbreviation, Si. Made by fusing crude silica with carbonate of soda; dissolve the residue, filter, and precipitate by hydrochloric acid.

This product must be triturated as per biochemic process before using internally.

This salt is the surgeon of the human organism. Silica is found in the hair, skin, nails, periosteum, the membrane, covering and protecting the bone, the nerve sheath, called neurilemma, and a trace is found in bone tissue. The surgical qualities of silica lie in the fact that its particles are sharp cornered. A piece of quartz is a sample of the finer particles. Reduce silica to an impalpable powder and the microscope reveals the fact that the molecules are still pointed and jagged like a large piece of quartz rock. In all cases, where it becomes necessary that decaying organic matter be discharged from any part of the body by the process of suppuration, these sharp pointed particles are pushed forward by the marvelous intelligence that operates without ceasing, day and night in the wondrous human Beth, and like a lancet cuts a passage to the surface for the discharge of pus. Nowhere in all the records of physiology or biological research can anything be found more wonderful than the chemical and mechanical operation of this Divine artisan.

The bone covering is made strong and firm by silica. In case of boils or carbuncle, the biochemist loses no time searching for "anthrax baccili" or germs, nor does he experiment with imaginary germ-killing serum, but simply furnishes nature with tools with which the necessary work may be accomplished.

## The Chemistry and Wonders of the Human Body

The Centaur of mythology is known in the "Circles of Beasts that worship before the Lord (Sun) day and night," as Sagittarius, the Archer, with drawn bow. Arrow heads are composed of flint, decarbonized white pebble or quartz. Thus we see why silica is the special birth salt of all born in the Sagittarius sign. Silica gives the glossy finish to hair and nails. A stalk of corn or straw of wheat, oats or barley would not stand upright except they contained this mineral.

Sagittarius people are generally swift and strong; and they are prophetic—look deeply into the future and hit the mark like the archer. A noted astrologer once said: "Never lay a wager with one born with the Sun in Sagittarius or with Sagittarius rising in the east lest you lose your wealth."

The Sagittarius native is very successful in thought transference. He (or she) can concentrate on a brain, miles distant ,and so vibrate the aerial wires that fill space that the molecular intelligence of those finely attuned to nature's harmonies may read the message.

Governing planet—Jupiter.

Gems—Carbuncle, diamond and turquoise.

The astral colors are gold, red and green.

Sagittarius is a fire sign and is represented in Bible alchemy by Levi, the third son of Jacob, meaning "joined or associated."

In the symbolism of New Testament Sagittarius corresponds with the disciple James, son of Alpheus.

# CAPRICORN: THE GOAT OF THE ZODIAC

### December 21 to January 19

CIRCLE means Sacrifice, according to the Cabala, the straight line bending to form a circle.

Thus we find twelve Zodiacal signs sacrificing to the sun. Symbolized by the devotions and sacrifices of the twelve disciples to Jesus.

Twelve months sacrifice for a solar year.

Twelve functions of man's body sacrifice for the temple, Beth or "Church of God"—the human house of flesh.

Twelve minerals—known as cell-salts—sacrifice by opertion and combining to build tissue.

The dynamic force of these vitalized workmen constitute the chemical affinities—the positive and negative poles of mineral expression.

The Cabalistic numerical value of the letters g, o, a, t, add up 12.

Very ancient allegories depict a goat bearing the sins of Israelites into the Wilderness.

In the secret mysteries of initiation into certain societies, the goat is the chief symbol.

In Alchemical lore the "Great Work" is commenced "in the Goat" and is finished in the "White Stone." Biochemistry is the "Stone the builders rejected" and furnishes the key to all the mysteries and occultism of the Allegorical Goat.

White stone also refers to the redemptive seed born in the Solar plexus known as Bethlehem, in Hebrew, or House of Bread, every 29½ days.

Those persons born between the dates December 21 and January 19 come under the influence of the Sun in Capricorn—the Goat. Capricorn represents the great business in-

## The Chemistry and Wonders of the Human Body

terests—trust and syndicates—where many laborers are employed. Thus Capricorn symbols the foundation and framework of society—the commonwealth of human interests.

The bones of the human organism represent the foundation stones and frame-work of the soul's temple (soul of man's temple).

See Solomon's Temple. Bone tissue is composed principally of the phosphate of lime, known as calcarea phosphate or calcium phosphate. Without a proper amount of lime no bone can be formed, and bone is the foundation of the body.

A building must first have a foundation before the structure can be reared. Thus we see why the "Great Work" commences in the Goat. Lime is white—hence the "White Stone."

In the 2nd chapter and 17th verse of Revelation may be found the alchemical formula of the "White Stone."

"To him that overcometh will I give to eat of the hidden manna, and will give him a White Stone, and in the Stone a new name written which no man knoweth saving he that receiveth it."

In the mountains of India, it is said, a tribe dwells, the priests of which claim that man's complete history from birth to death is recorded in his bones. These people say the bones are secret archives, hence do not decay quickly as does flesh and blood.

When the molecules of lime phosphate fall below the standard, a disturbance often occurs in the bone tissue and the decay of bone, known as *carries* of bone, decay of bone commences. Phosphate of lime is the worker in albumen. It carries it to bone and uses it as cement in the making of bone.

So-called Bright's disease (first discovered in a man named Bright) is simply an outflow of albumen via kidneys, due to a deficiency of phosphate of lime.

When the Goat Salt is deficient in the Gastric Juice and bile ferments arise from undigested foods, acids are formed and which find their way to syno-vial fluids in the joints of

## Capricorn: The Goat of the Zodiac

legs or arms or hands and often cause severe pains, but why the chemical operation, which is perfectly natural, should be called rheumatism passeth understanding.

Non-functional albumen, caused by a lack of lime phosphate, is the cause of eruptions, abscesses, consumption, catarrh and many other so-called diseases.

But let us all remember that disease means not-at-ease, and that the words do not mean an entity of any kind, shape, size, weight or quality, but an effect caused by some deficiency of blood material.

Phosphate of lime should never be taken in crude form. It must be triturated to 6th x, according to the biochemic method, in milk sugar in order to be taken up by the mucous membrane absorbents, and thus carried into the circulation.

Capricorn people possess a deep interior nature in which they often dwell in the "Solitude of the Soul."

They scheme and plan and build air castles and really enjoy their ideal world. If they are sometimes, talkative, their language seldom gives any hint of the wonderland of their imagination.

To that enchanted garden the sign, "No Thoroughfare," forever blocks the way.

The Capricorn gems are white Onyx and Moonstone. The astral colors are garnet, brown, silver-gray and black.

Capricorn is an earth sign.

In Bible alchemy, Capricorn represents Judah, the fourth son of Jacob, and means "the praise of the Lord." In the symbolism of the New Testament, Capricorn corresponds with the disciple John.

# THE SIGN OF THE SON OF MAN: AQUARIUS

### January 20 to February 19

O age of man: Aquarius,
   Transmuter of all things base,
"Son of Man in the Heavens,"
   With sun-illumined face."

Our journey was long and weary,
   With pain and sorrow and tears,
But now at rest in thy kingdom,
   We welcome the coming years.

THOSE born between the dates January 20 and February 19 are doubly blest, and babies to be born during that period for many years to come will be favored of the gods.

The Solar System has entered the "Sign of the Son of Man," Aquarius, where it will remain for over 2,000 years. According to planetary revolutions the Sun passes through Aquarius once every solar year; thus we have the double influence of the Aquarius vibration from January 20 to February 19.

Air contains 78 per cent of nitrogen gas, believed by scientists to be mineral in ultimate potency. Minerals are formed by the precipitation of nitrogen gas. Differentiation is attained by the proportion of oxygen and aqueous vapor (hydrogen) that unites with nitrogen.

A combination of sodium and chlorine forms the mineral known as common salt. This mineral absorbs water. The circulation or distribution of water in the human organism is due to the chemical action of the molecules of sodium chloride.

Crude soda cannot be taken up by mucous membrane absorbents and carried into the circulation. The sodium

## Aquarius: The Sign of the Son of Man

molecules found in the blood have been received from vegetable tissue which drew these salts from the soil in high potency. The mineral, or cell-salts, can also be prepared (and are prepared) in biochemic or homeopathic potency as fine as the trituration of Nature's laboratory in the physiology of plant growth, and then they are thoroughly mixed with sugar of milk and pressed into tablets ready to be taken internally to supply deficiencies in the human organism. A lack of the proper amount of these basic mineral salts (twelve in number) are the cause of all so-called disease.

Common table salt does not enter the blood, being too coarse to enter the delicate tubes of mucous membrane absorbents, but this salt does distribute water along the intestinal tract.

Aquarius is known in astrological symbol as "The Water Bearer." Sodium chloride, known also as natrum muriaticum, is also a bearer of water, and chemically corresponds with the zodiacal angle of Aquarius.

The term angle, or angel, of the Sun may also be used, for the position of the Sun at birth largely controls the vibration of blood.

So, then, we have sodium chloride as the "birth salt" of Aquarius people.

The governing planets are Saturn and Uranus; the gems are sapphire, opal and turquoise; the astral colors are blue, pink and nile green. Aquarius is an air-sign.

In Bible Alchemy, Aquarius represents Dan, the fifth son of Jacob, and means "judgment," or "he that judges." In the symbolism of the New Testament, Aquarius corresponds with the disciple James.

Uranus, the revolutionary planet, known as the "Son of Heaven," is now in Pisces, "Sign of the Fish," and will remain there until January, 1927.

# PISCES: THE FISH THAT SWIM IN THE PURE SEA
## February 19 to March 20

MOST everybody knows that Pisces means fishes, but few there be who know the esoteric meaning of fish. Fish in Greek is Ichthus, which Greek scholars claim means "substance from the sea."

Jesus is derived from the Greek for fish; Mary, mare, means water; therefore we see how the Virgin Mary, pure sea, gives birth to Jesus, or fish. There are two things in the universe—Jesus and the Virgin Mary—spirit and substance. So much for the symbol or allegory.

From the earth viewpoint we say that the Sun enters the Zodiacal sign Pisces February 19, and remains until March 21. This position of the Sun at birth gives the native a kind, loving nature, industrious, methodical, logical and mathematical; sympathetic and kind to people in distress.

In the alchemy of the Bible we find that the sixth son of Jacob, Naphtali, which means "wrestling of God," symbols Pisces, for the Pisces native worry and fret because they cannot do more for their friends or those in trouble.

The phosphate of iron is one of the cell-salts of human blood and tissue. This mineral has an affinity for oxygen which is carried into the circulation and diffused throughout the organism by the chemical force of this inorganic salt. The feet are the foundation of the body. Iron is the foundation of blood. Most diseases of Pisces people commence with symptoms indicating a deficiency of iron molecules in the blood; hence it is inferred that those born between the dates of February 19 and March 21 use more iron than do those born in other signs.

Iron is known as the magnetic mineral, due to the fact that it attracts oxygen. Pisces people possess great magnetic force in their hands and make the best magnetic healers.

## Pisces

Health depends upon a proper amount of iron phosphate molecules in the blood. When these oxygen carriers are deficient, the circulation is increased in order to conduct a sufficient amount of oxygen to the extremities—all parts of the body—with the diminished quantity of iron on hand. This increased motion of blood causes friction, the result of which is heat. Just why this heat is called fever is a conundrum; maybe because fever is from Latin fevre, "to boil out," but I fail to see any relevancy between a lack of phosphate of iron and "boiling out."

The phosphate of iron (ferrum phosphate), in order to be made available as a remedy for the blood, must be triturated according to the biochemic method with milk sugar up to the third or sixth potency in order that the mucus membrane absorbents may take it up and carry it into the blood. Iron in the crude state, like the tincture, does not enter the circulation, but passes off with the faeces and is often an injury to the intestinal mucous membrane.

The governing planet of this sign is Jupiter.

The gems are chrysolite, pink-shell and moonstone.

The astral colors are white, pink, emerald-green and black.

Pisces is a water sign.

In Bible alchemy, Pisces represents Naphtali, the sixth son of Jacob, and means "wrestlings of God." In the symbolism of the New Testament, Pisces corresponds with the disciple Philip.

The birth of Benjamin is given in that wonderful allegory, the 35th chapter of Genesis.

Benjamin is therefore the 13th child of Jacob.

See article on "13" in the great book "God-man: The Word Made Flesh."

> "The heavens declare the glory of God;
> And the firmament showeth his handiwork.
> Day unto day uttereth speech,
> And night unto night showeth knowledge.
> There is no language where
> There voice is not heard."
> —Psalms, 19th Chapt., 1-5 v.

# BIOPLASMA, LIFE SUBSTANCE

IT is a combination of the twelve cell-salts of the human body in the exact proportion contained in healthy blood.

Bioplasma is a perfect nerve, brain and blood food, composed of the phosphates, sulphates and chlorides contained in vegetables, fruits, grains and nuts.

This combination of the inorganic cell-salts raises the chemical vibration of blood that has become depleted by a deficiency of the necessary workers.

A sufficient amount of the cell-salts of the body, properly combined and taken as food forms blood that materializes in healthy fluids, flesh and bone tissue.

We should take the tissue cell-salts as one uses health foods, not simply to change not health to health, but to keep the rate of blood vibration in the tone of health all the time.

In the new age, we will need perfect bodies to correspond with higher vibration, or motion of the new blood, for "old bottles (bodies) cannot contain the new wine."

We have many letters from people of all classes who have been cured by Bioplasma after trying all other remedies and methods, without relief. We do not publish cures or give names, except when permission is given in writing. Those cured by this treatment advertise the fact in a more decent and effective way than by having their names and ailments made merchandise of for commercial purposes.

Each Sign of the Zodiac is represented by the twelve functions of the body, and the position of Sun at birth.

Therefore, the cell-salt corresponding with the Sign of Zodiac and function of body is consumed more rapidly than other salts and needs an extra amount to supply the deficiency caused by Sun's effect, therefore always give date of birth when ordering Bioplasma.

This treatment should be continued for one year, thus supplying the blood while the Sun passes through the twelve Zodiacal Signs.

# DIET

DIAGNOSIS unnecessary as *symptoms* of so-called disease are results of a cause, and the cause of the symptoms is *deficiencies*—not entities—but *lack of entities.*

Crude salt, and manufactured sugar, are practically unnecessary in the human organism. All food stuffs contain sodium chloride (common table salt), in high potency. Food also contains the twelve cell-salts—sodium, lime, potash, iron, magnesia and silica, but we only use them in biochemic therapeutics in potency or fineness, corresponding to their presence in foods, grains, nuts, vegetables, etc., in order that they may be taken up by mucous membrane absorbents and thus enter into the blood vessels.

Common salt, as a condiment, must go to the island of banishment with crude sugar and poisonous drugs, including the deadly nicotine. All food stuffs contain natural sugar. A reasonable amount of honey is beneficial.

Over-eating is the supreme curse of the race. All food that is not completely *digested* (burned up as fuel), decays and ferments and produces *acids. All diseases originate in the intestinal tract.* The law has prohibited the commercial distillery; let men and women prohibit the distilleries *in their own bodies* by governing their appetites, and thus work out their own salvation.

Flesh eating is abhorent to the spiritual man.

*Ani-mal* means "bad life," that is, negative, or bad by contrast with spiritual life. If we should say ox flesh instead of beef; hog flesh instead of pork, or ham; or corpse instead of meat, there would be no need of prosecuting the meat packers, for they would go bankrupt for lack of patronage.

# THE BRIDGE OF LIFE
"Conceived in sin and brought forth in iniquity."

"A NOISELESS, patient spider,
I mark'd, where, on a little promontory, it stood, isolated;
Mark'd how, to explore the vacant, vast surrounding,
It launch'd forth filament, filament, filament, out of itself;
Ever unreeling them—ever tirelessly speeding them.

"And you, O my soul, where you stand,
Surrounded, surrounded, in measureless oceans of space,
Ceaselessly musing, venturing, throwing—seeking the spheres, to connect them;
Till the bridge you will need, be form'd—till the ductile anchor hold;
Till the gossamer thread you fling, catch somewhere, O my soul."
—*Walt Whitman.*

"O Man of Earth, watch well the steps thou findest,
Spread out before thy feet by cosmic plan;
Do thy soul's best, with body and with mind,
To pay thy debt, and bridge this Karmic Span."
—*Edith F. A. U. Painton.*

The statement of Holy Writ, that "man is conceived in sin and brought forth in iniquity" has a three-fold meaning, viz., chemical, physiological and astrological. The real meaning in the original is, that the human embryo remains nine months in the female laboratory, thus falling short three months of completing a solar or soul year. It also represents the journey of the ego from the moon to earth, or conception. Twelve, which represents the circle and stands for completion.

The word sin comes from Schin, the twenty-first letter of the Hebrew alphabet, and means to fall short of completeness, or understanding, wisdom. In the Tarot symbol, S, or Sin, is represented by the "Blind Fool," one lacking in wisdom, "Brought forth in iniquity" is merely a repetition of the words "born in sin." Iniquity and inequity or unequal, mean the same. The ancient Hebrews called Moon, Sin, because it gave light only part of the time.

To acquire wisdom that will enable the Ego in flesh to build a bridge across the three-month gap, or space between the point of conception and birth, is the one real problem that confronts the ego on the material plane of expression. The alchemists, seers and astrologians of all ages have wrestled with this problem in their ceaseless endeavors to unravel the great mystery of man's dominion over flesh. Whether it be the chemist seeking new compounds, the physiologist searching and testing the fluids of the fearfully and wonderfully made body of man, the alchemist probing for the Elixir of Life—the Ichor of the Gods or the astrologian pulling and adjusting the etheric wires that criss-cross the spaces in an earnest desire to make good and sane the statement "The wise man rules his stars,"—all, all are seeking to span the awful space that yawns between the neophyte and the Promised Land of immortality in the body, where "in my flesh I shall see God," and when and where he can truly say with the regenerated Job, "I have heard of thee by the hearing of the ear, but now mine eye seeth thee." Man must work out his own salvation.

The bridge to be built across the three-months space must have a mineral base or rock foundation. "Thou art Peter (petra, stone, or mineral), on thee will I build my church," etc. Church is from the second Hebrew letter, Beth, a house temple, or church. The human body is a house, temple, or church for the Soul which may be lost or saved by the higher self or spiritual ego residing in the cerebellum the "Secret Place of the Most High." "Know ye not that your bodies are the temple (church) of God?"

There are twelve inorganic mineral cell-salts in the human body, and these minerals (stones in the temple) correspond in vibration to the twelve signs of the Zodiac. During the nine months of gestation the embryo receives and appropriates the creative energies of nine of these salts, leaving three to be supplied after the parting of the umbilical cord. Take for example a native born Febru-

## The Chemistry and Wonders of the Human Body

ary 22nd, with the Sun's entry into Pisces: The embryo, having begun its journey at the gate of Gemini and negotiated the nine gestatory signs, his blood vibration at birth is thus deficient in the qualities of Pisces, Aries and Taurus, as also in the chemical dynamics of phosphate of iron, phosphate of potassium and sulphate of sodium—the mineral bases respectively of the signs of this incompleted quadrant. In so far as his circulatory system may receive these needed builders, the health will be balanced and life prolonged.

The chemical union of these cell-salts with organic matter, such as oil, fibrin, albumen, etc., forms the various tissues of the body and administers to the physiological needs as represented by the Bridge, that the multiple cells may respond more harmoniously and completely to the magic touch of the Divine energy, just as the tones of a musical instrument are made the more melodious through a properly skilled manipulation. And as bridge-building in a mechanical sense depends upon the plans and specifications of a competent civil engineer, so does the Bridge of Life depend upon the astrologian to chart and compass the way.

Our diagram indicates at a glance the chemical formula that pertain respectively to the zodiacal divisions, but to give a clearer conception as regards their specific qualities and physiological action in relation to the various signs, reference may be had to the following compendium:

The coming of Christ and the end of the world has been preached from every street corner for several years, and thousands are pledging themselves to try to live as Christ lived or according to their concept of His life.

No great movement of the people ever occurs without a scientific cause.

The Optic Thalamus, meaning "light of the chamber," is the inner or third eye, situated in the center of the head. It connects the pineal gland and the pituitary body. The optic nerve starts from this "eye single." "If thine eye be single, thy whole body will be full of light." The *optic thalamus is the Aries planet* and when fully developed through physical regeneration it lifts the initiate up from the Kingdom of Earth, animal desire below

## The Bridge of Life

the solar plexus, to the pineal gland that connects the cerebellum, the temple of the Spiritual Ego, with the optic thalamus, the third eye.

By this regenerative process millions of dormant cells of the brain are resurrected and set in operation, and then man no longer "sees through a glass darkly," but with the Eye of spiritual understanding.

We venture to predict that the planet corresponding to the optic thalamus will soon be located in the heavens.

"The new order cometh." Mars must be dethroned as ruler of the brain of man.

To those who object to linking chemistry with astrology, the writer has this to say:

The Cosmic Law is not in the least disturbed by negative statements of the ignorant individual. Those investigators of natural phenomena, who delve deeply to find Truth, pay little heed to the babbler who says, "I can't understand how the zodiacal signs can have any relation to the cell-salts of the human body." The sole reason that he "cannot understand" is because he never tried to understand.

A little earnest, patient study will open the understanding of any one possessed of ordinary intelligence and make plain the great truth that the *UNI*verse is what the word implies, i.e., *one verse*.

It logically follows that all parts of one thing are susceptible to the operation of any part.

The human body is an epitome of the cosmos.

Each sign of the Zodiac is represented by the twelve functions of the body and the position of the Sun at birth.

Therefore, the cell-salt corresponding with the Sign of the Zodiac and function of the body is consumed more rapidly than other salts and needs an extra amount to supply the deficiency caused by the Sun's influence at that particular time.

In ancient lore Aries was known as the "Lamb of Gad," or God, which represents the head or brain. The brain controls and directs the body and mind of man. The brain itself, however, is a receiver operated upon by celestial influences or angles (angels) and must operate

[ 153 ]

## The Chemistry and Wonders of the Human Body

according to the directing force or intelligence of its source of power.

Man has been deficient in understanding because his brain receiver did not vibrate to certain subtle influences. The dynamic cells in the gray matter of the nerves were not finely attuned and did not respond—hence sin, or falling short of understanding.

From the teachings of the Chemistry of Life we find that the basis of the brain or nerve fluid is a certain mineral salt known as potassium phosphate, or Kali Phos.

A deficiency in this brain constituent means "sin," or a falling short of judgment or proper comprehension. With the advent of the Aries Lord, God, or planet, cell-salts are rapidly coming to the fore as the basis of all healing. Kali phosphate is the greatest healing agent known to man, because it is the chemical base of material expression and understanding.

The cell-salts of the human organism are now being prepared for use, while poisonous drugs are being discarded everywhere. Kali phosphate is the especial birth salt for those born between March 21 and April 19.

These people are brain workers, earnest, executive and determined—thus do they rapidly use up the brain vitalizers.

The Aries gems are amethyst and diamond.

In Bible alchemy Aries represents Gad, the seventh son of Jacob, and means "armed and prepared"—thus it is said when in trouble or danger, "keep your head."

In the symbolism of the New Testament, Aries corresponds with the disciple Thomas. Aries people are natural doubters until they figure a thing out for themselves.

The astronomer, by the unerring law of mathematics applied to space, proportion, and the so far discovered wheels and cogs of the uni-machine, can tell where a certain planet must be located before the telescope has verified the prediction. So the astro-biochemist knows there must of necessity be a blood mineral and tissue builder to correspond with each of the duodenary segments that constitutes the circle of the Zodiac.

## The Bridge of Life

Not through quarantine, nor disinfectants, nor boards of health, will man reach the long-sought plane of physical well-being; nor by denials of disease will bodily regeneration be wrought; nor by dieting or fasting or "Fletcherizing" or suggesting, will the Elixir of Life and the Philosopher's Stone be found. The Mercury of the Sages and the "hidden manna" are not constituents of health foods. Victims of salt baths and massage are bald before their time, and the alcohol, steam and Turkish bath fiends die young. Only when man's body is made chemically perfect will the mind be able perfectly to express itself.

And the secret of this chemical perfectionment is the sum total of the requirements involved in this zodiacal Bridge. The rock—Peter, or Petra—must be completed before the etheric wires that span the gulf between birth forward to the sidereal point of conception can vibrate in such harmony as to sustain the traveller on this "magical bridge of three piers," or the three zodiacal signs through which the material body must successfully function before it may hope to lift the veil of Isis.

The Bridge of Life, a symbol of physical re-genesis, has been exploited in song, drama, and story. Paracelsus, Pythagoras, Lycurgus, Valentin, Wagner, and a long and unbroken line of the Illuminati, from time immemorial have chanted their epics in unison with this "riddle of the Sphinx," across the scroll of which is written, "Solve me, or die."

Of all the multiple adepts or masters that have kept the lights burning above the Three Piers of the magical Bridge, none has more clearly and beautifully written thereof than did the great astrologian poet, Isaiah:

"Then the eyes of the blind shall be opened, and the ears of the deaf shall be unstopped. Then shall the lame man leap as a hart, and the tongue of the dumb shall sing; for in the wilderness shall waters break out, and streams in the desert. And the glowing sand shall become a pool, and the thirsty ground springs of water; in the habitation of jackals, where they lay, shall be grass with reeds and rushes. And a *highway shall be there, and a way and it shall be called, The way of holiness;* the un-

## The Chemistry and Wonders of the Human Body

clean shall not *pass over it,* but it shall be for the redeemed; the wayfaring men, yet fools, shall not err therein."

Here we have the last step on the physical-plane that breaks down the "middle wall of partition."—*Paul.* Then the Ego is enabled to regenerate by saving the Word of God—the Seed—and thus render further incarnations unnecessary.

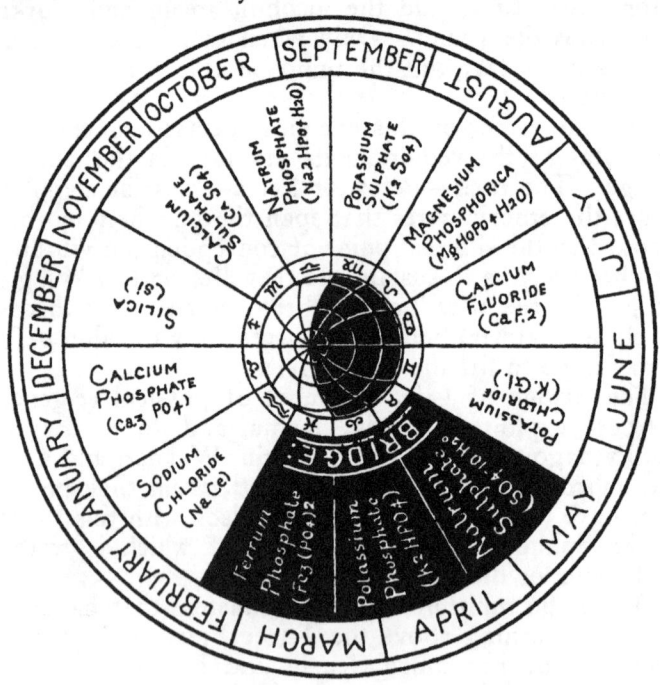

CPSIA information can be obtained
at www.ICGtesting.com
Printed in the USA
LVHW091543120120
643230LV00005B/493/P

9 781614 275121